The Writing System Workbook

A step-by-step guide for business and technical writers

Judith H. Graham, Ph.D., and Daniel O. Graham, Jr.

Preview Press—Fairfax, Virginia

ISBN 0-9644495-0-1

Printed in the United States of America

Other Books by the Grahams

The Gatekeepers — Baen Books; New York, NY; 1995 (Winner of the 1996 Compton Crook Award)

The Politics of Meaning — Preview Press; Fairfax, VA; 1995

Entering Tenebrea, with Roxann Dawson — Pocket Books; New York, NY; 1999

Preface

The writing system taught in this book serves as a best practice in large U.S. corporations. By using this system, you write faster. By using the system's techniques, you organize your thoughts better and get to the point quickly. You write clear, concise, better quality documents.

We encourage every professional to learn this writing system and practice its techniques. Today, much corporate product is written communication. Increased writing speed cuts cost, and improved document quality increases client satisfaction. Therefore, professionals who use the writing system increase profits.

The Writing System Workbook's authors, Daniel and Judith Graham, Ph.D., teach writing seminars and provide one-on-one writing coaching throughout the United States and abroad.

Acknowledgments

We thank the thousands of bright engineers, scientists, and business professionals who, during our writing seminars, gave us valuable feedback and helped us refine our writing techniques. This book is yours.

Contents

The Writing System Workbook

A step-by-step guide for business and technical writers

This workbook teaches you a system of proven techniques to help you write business and technical documents faster and better.

We wrote this workbook for practical, task-oriented engineers, scientists, and business professionals. If you've mastered your discipline, you can master these simple techniques. Think of this workbook as your user's manual for written communication with built-in tutorial. Use this workbook three ways:

1. Learn the writing system by doing the exercises—about 40 hours work.
2. Consult the step-by-step guide as you write your documents.
3. Refer to the book to solve writing problems.

Designed for specific how-to instruction, this workbook follows the writing system: 3 phases, 13 steps, and 43 techniques. Writing techniques feature tips, warnings, some error traps, examples, references, brief—emphasize *brief*—discussions, and exercises. We put the answers to the exercises in Appendix A.

Relax. Set your own pace. And enjoy taking the mystery out of good writing.

Overview of the Writing System

Follow this system to write more quickly and get better results.

The Writing System has three phases. Within each phase you must complete certain steps. For each step, we provide simple techniques.

We call the three phases **pre-writing**, **writing,** and **post-writing:**

- Pre-writing Phase—Budget 25% of your time to accomplish five steps to prepare for composing the draft.

- Writing Phase—Budget 25% of your time to compose the draft.

- Post-writing Phase—Budget 50% of your time to accomplish seven steps to craft the draft into a quality document.

By following this rigorous system, you get tangible benefits:

1. You manage time. You'll know how much time to budget to each writing phase, even to each writing step, and you can monitor your progress to meet deadlines.

2. You increase writing speed. Writers waste much time when they do not know what to do next, or lack clear techniques to solve writing problems.

3. You improve the quality of your documents. The pre-writing steps ensure that you present your information the way the audience wants to see it. The post-writing steps ensure quality in content, style, and grammar.

The Writing System

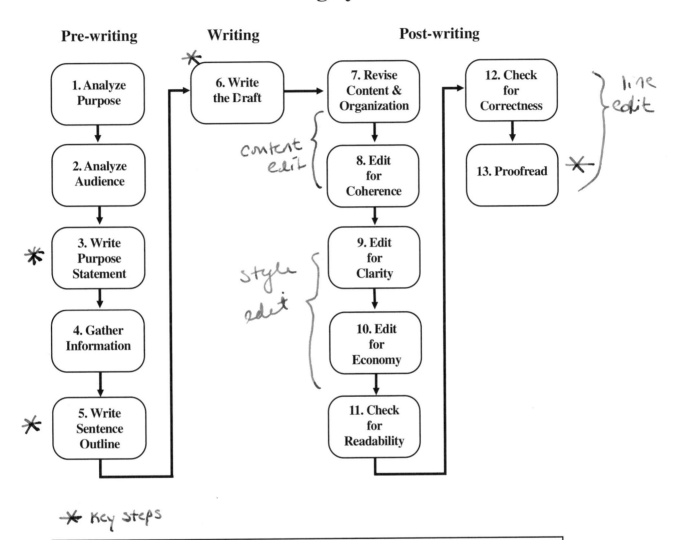

Pre-writing   **Writing**   **Post-writing**

1. Analyze Purpose

2. Analyze Audience

3. Write Purpose Statement

4. Gather Information

5. Write Sentence Outline

6. Write the Draft

7. Revise Content & Organization

8. Edit for Coherence

9. Edit for Clarity

10. Edit for Economy

11. Check for Readability

12. Check for Correctness

13. Proofread

content edit

style edit

line edit

✱ Key steps

Discussion:

Our system imposes discipline over the messy business of writing. Our phases, steps, and techniques help you write your document faster. Ideally, you can make one systematic pass at a document.

You use different skill sets during each writing phase. During the pre-writing phase, you use your analytical skills. During the writing phase, you use your composing skills. During post-writing, you use your editing skills.

However, longer documents or collaborative writing projects do not conform neatly to a single, never-look-back pass, because you embrace new ideas or discover gaps while writing and editing the document. Our system helps here as well. Many techniques in our system act as error traps to catch problems and expose gaps early. Fortunately, each error trap points directly back to a former step and suggests the technique you can revisit to fix the problem.

Pre-writing Phase

The Pre-writing Phase includes all the preparation needed *before* you write your draft. Use your analytical skills and follow these five steps to prepare well.

Step 1   **Analyze Purpose**

Step 2   **Analyze Audience**

Step 3   **Write Purpose Statement**

Step 4   **Gather Information**

Step 5   **Write Sentence Outline**

Discussion:

During the pre-writing phase, you use your analytical—problem-solving—skills. As in other endeavors, you must finish your analysis before execution. Or in our case, you must finish pre-writing before writing the draft.

In most projects, a strong beginning ensures smooth execution and a satisfactory result: the same holds true for writing. Do not neglect pre-writing. Skipping pre-writing invariably slows you down and causes problems in content, organization, and style.

Step 1. Analyze Purpose

1. Analyze Purpose

List and contrast purposes

Begin your writing task by analyzing purpose using this technique:

1.1 List and contrast three purposes associated with your writing task.

Discussion:

Some documents serve no purpose for these reasons:

- The message may be worthwhile, but doesn't need to be written.
- The message may be worthwhile, but the author fails to tell the reader why.
- The message is not worthwhile.

Don't waste time by writing documents that lack purpose. Specifically, don't hide behind impersonal memos to avoid confrontations. You may document a conflict that might be better forgotten. Also, don't write anything you would not want read back to you in court.

Technique 1. 1	**List and contrast three purposes associated with your writing task: your purpose, the reader's purpose, and the purpose of the work.**

Tips: Consider the following:
Why are you writing?
Why is the reader reading?

Warning: Do not confuse the purpose of the work with your purpose for writing about it.

Example: Purposes associated with writing a proposal to develop order-fulfillment software to a mail-order business

- Your purpose for writing: *Win the contract*
- Reader's purpose for reading: *Select a vendor to solve the order-fulfillment problem*
- Purpose of the work: *Develop order-fulfillment software*

Purposes associated with writing a letter of reprimand to an employee who submitted a false time sheet

- Your purpose for writing: *Record infraction and discipline employee*
- Reader's purpose for reading: *Learn how to avoid losing job*
- Purpose of the work: *To account for billable hours*

See also: Purpose Statement.

Discussion:

As you define your purpose for writing and your reader's purpose for reading, consider these advantages of written communication:

- accuracy—you can provide many details with precision
- economy—you can reach a broad audience with the same message
- record—you can provide an audit trail or record for future readers

Your document succeeds when it satisfies both your purpose for writing and your reader's purpose for reading. However, many writers mistakenly focus solely on the purpose of the work and neglect the more important purposes associated with the writing task. Include the work's purpose in your background material.

Exercise: List and contrast three purposes associated with each writing task. (Answer A-1)

1. You must write a required monthly status report to your client, the City of Boston, who hired your company to clean the water in Boston Harbor.

- Your purpose for writing:

- Reader's purpose for reading:

- Purpose of the work:

2. After receiving many phone queries, you decide to write a memo to your plant employees describing procedures for getting their monthly parking sticker from their shift supervisor.

- Your purpose for writing:

- Reader's purpose for reading:

- Purpose of the work:

3. As office manager, you write a staff study to your immediate supervisor recommending leasing rather than buying a copy machine.

- Your purpose for writing:

- Reader's purpose for reading:

- Purpose of the work:

Final exercise: List and contrast three purposes associated with this writing task. (Answer A-1)

Scenario: You are a regional sales manager for a rapidly expanding chain of hardware stores, 20 stores in your region. You communicate directly to the store managers concerning new products, price changes, and promotional campaigns to ensure that the store managers know how best to stock their shelves and take advantage of promotions to boost store sales.

You and the store managers get rewarded well for exceeding sales goals; consequently, all are motivated to boost sales.

Until now, you've barraged the store managers with letters, memos, telexes, and faxes—often passing on any brochures or other materials provided by manufacturers. However, far from grateful for your detailed communications, the store managers complain that they haven't time to read all your "stuff." They claim your varied missives get lost in the other piles of communication they get from headquarters. And when they do read your "stuff," they can't easily figure out exactly what you want them to do, so they ignore your message. If they do figure out your message, they delegate the execution of your instructions to floor clerks.

You decide to streamline your written communications with the store managers to capture their attention, transmit your message, and boost sales.

- Your purpose for writing:

- Store manager's purpose for reading:

- Purpose of the work:

Step 2. Analyze Audience

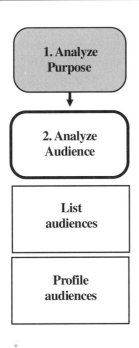

After analyzing purposes associated with your writing task, use two techniques to analyze your audience:

2.1 List your audiences in priority order. Identify each as expert, manager, operator, or layperson.

2.2 Profile each audience using a checklist.

Discussion:

A reader is a person. An audience is one or more readers with common interests and background. In simpler documents, like letters and memos, the reader and audience is often one person. In more involved documents like plans or proposals, you may have to deal with many readers whom you must group into one or more audiences.

The most common failing in business and technical writing is a poor sense of audience. Writers, especially subject-matter experts, too often dive straight into writing about what they know. They end up with documents written for themselves.

Before you write, you need to determine *why* the audience reads your document: to simply learn more about your subject, to make a decision, to perform a task?

Then you must determine *how* the audience wants to see your information. Does your audience want you to summarize some information or spell it out? How much information does the audience want and need? Is your audience mostly interested in conclusions and recommendations, or does your audience want a full discussion of your findings and methods as well? Does your audience want visuals to make certain information more accessible?

Technique 2.1	**List your audiences in priority order. Identify each as expert, manager, operator, or layperson.**

Tips: Determine who the readers are: one or many.
Group readers into a primary (perhaps secondary, and tertiary) audience. Your primary audience is the most important audience for your document.

Determine how each audience uses the document:

- to stay informed—Expert
- to make a decision—Manager
- to perform a task—Operator
- to get general information—Layperson

Warning: Don't assume that audiences' jobs or background dictate what they want from your document. Vice President Lee may want to perform a task; Doctor Durant may want to make a decision.

Example:

Audience	**Ms. Jones**	**Dr. Albert**	**Code Writers**
Priority	*Primary*	*Secondary*	*Tertiary*
Reading as	Manager	Expert	Operators

See also: Coherence.

Discussion:

You'll deal with four kinds of audience for business and technical writing: expert, manager, operator, or lay person.

Expert—wants to follow your steps to replicate your reasoning and arrive at your same conclusion.

Manager—only reads enough of the document to make a decision. Wants the conclusions first, then the discussion, and the data last, if at all.

Operator—wants just enough to know how to complete the task, no theory. Wants precise details, including pictures.

Lay person—wants practical information, less theory. Wants less jargon, more examples and analogies.

Note this subtle distinction: You identify your audience by how readers use the document. For example, your readers may be computer experts, but they use your document to decide which computer to purchase. Because your audience reads to make a decision, you write to the manager—not the expert—audience.

Exercise: Fill in the blank to identify the audience matching each set of characteristics. (Answer A-2)

1._____

This audience needs precision. They need to know what and when more than why. They want to know—What procedures must I follow to begin, continue, and complete a job? What equipment must I install or operate? What does it look like and how does it work? Are alternate procedures useful or possible? What can go wrong, and how do I correct if it does?

2._____

This audience wants the interesting or useful nuggets of information. They read for curiosity or self-interest. They won't read technical language and theories. They consider many details unnecessary. They want you to translate jargon into standard English, using short words and short sentences when you discuss difficult subjects. They need analogies and examples to understand your message. They need just enough background to follow your discussion easily.

3._____

This audience wants a concise presentation of practical information. They want to know—Is this proposal feasible? What will it cost? Is it cost effective? Who is doing it? How well is it being done? Are things running on schedule? What's going to happen next? What advantages will result for the organization? What do you want me to do?

4._____

This audience requires detail. They may want to know why as well as what. They want to know—What problems need to be solved? What procedures will be used? Does the product meet design and performance specifications? Is there adequate and documented data to support conclusions? How can these conclusions help us in further research, product development, or training?

Technique 2.2   Profile each audience using a checklist.

Tips:	Determine what they know, what they need to know, and how much time they have to learn and use the information.
	If you don't know, you must assume.
Warning:	Don't over-estimate your audience's knowledge, time, or patience.
Example:	Profile the three audiences for a document about International Electronic Funds Transfers.

Audience	**Ms. Jones**	**Dr. Albert**	**Code Writers**
Priority	*Primary*	*Secondary*	*Tertiary*
Reading as	Manager	Expert	Operators
Profession	Finance	Computer	Programming
Knowledge of subject	General	Technical	None
Friendly, hostile, neutral friendly	neutral	neutral	friendly
Considers you expert	yes	yes	yes
Wants background	yes	no	no
Wants theory	no	no	no
Wants examples, pictures	no	yes	yes

See also:	Purpose statement.

Discussion:

As you profile your audiences, remember each criteria:

- Profession, knows subject—tells you what level of detail you need, how much jargon you can use, and what you can assume the audience knows.
- Friendly, considers you expert—helps you determine tone, whether you write informally, persuasively, or authoritatively. The audience's opinion of you also affects your content. Do you have credibility, or must you back up everything to "prove" your case?
- Wants background, theory, examples—ties directly to how audiences read: manager, expert, operator, layperson. Audience expectations affect your content and suggest their tolerance for technical language.

Remember that most of us overestimate the reader's familiarity with our subject—when in doubt, spell it out.

Exercise: List and profile audiences. (Answer A-2)

Scenario: You are a regional sales manager for a rapidly expanding chain of hardware stores. You have 20 stores in your region, each with a store manager. Most store managers have more than ten years' experience with your company, and therefore have a detailed knowledge of store operations.

You decide to send a monthly marketing summary to the 20 store managers informing them about new products, price changes, and promotional campaigns. You also intend to tell them how best to stock their shelves, and set up displays to take advantage of promotions to boost store sales.

You know that most store managers delegate the execution of your instructions to floor clerks. The floor clerks typically have one to three years' experience displaying merchandise and serving customers. List and profile your audiences.

Audience	**Store Mgrs.**	**Floor Clerks**
Priority	*Primary*	*Secondary*
Reading as		
Profession		
Knowledge of subject		
Friendly, hostile, neutral		
Considers you expert		
Wants background		
Wants theory		
Wants examples, pictures		

Final exercise: This memo's writer skipped analyzing purpose and audience. Analyze purpose and audience, profile the audience, and recommend changes to the memo. (Answer A-3)

Interoffice Correspondence
Systems Evaluation Team

Subject: TaskPert Project Management Software
Date: 22 January 1994
To: J. Brown   VP-Strategic Planning
From: K. Smith   Analyst

 I recently completed an informal evaluation of the TaskPert Project Management Software. We had requested a one-month trial from BSH Software, Inc. with the understanding that we would purchase TaskPert if it met our requirements. Following is a list of positive and negative aspects of the TaskPert software package.

Positive:

1. Taskpert features three different kinds of milestones for recurring tasks.

2. It has a 200-character note field suitable for lengthy task descriptions.

3. It has "subnetworks," which we call "subschedules" to relate different jobs.

4. Taskpert can generate custom reports and save report forms for later use.

5. Taskpert seems to be the Project Managment software of choice for our industry.

6. BSH indicated they would discount the price.

Negative:

1. The program uses a confusing web of menus which I found cumbersome when changing functions while working among several files.

2. The documentation for the software is not well organized.

3. The software does not support our Calcucomp Instruments Plotter.

Although I was generally pleased with the program, the inability of the software to support our plotter eliminates the possibility of a switch to Taskpert. Therefore, we have returned the software, and we asked BSH Software to contact us if they add CI plotter support.

Step 3. Write a Purpose Statement

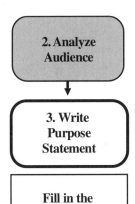

After analyzing your purpose for writing and your audience's purpose for reading, you can determine the *document's* purpose.

Use two techniques to write a purpose statement for your document:

3.1 Fill in the five parts and write the document's purpose statement: Actor, Action, Audience, Object, Outcome.

3.2 Use the purpose statement to focus yourself and others.

Discussion:

Many writers find themselves overwhelmed by their writing task. Why? Because their writing task is too broad. They try to deal with too many topics for too many audiences with too many purposes—many of which conflict. A well-crafted purpose statement limits your document by defining the

- format
- one action to be accomplished by the document
- primary audience
- one object of discussion
- reason why the audience should read your document

A well-crafted purpose statement manages your expectations for the document. Save your purpose statement. Later, when editing for coherence, use the purpose statement in an introduction to manage your *readers'* expectations.

Technique 3.1	**Fill in the five parts for a purpose statement: Actor, Action, Audience, Object, Outcome.**

Tips: Use your purpose and audience analysis to determine each of the five parts of the purpose statement. Evaluate each part to make sure they work together.

Actor =	the document itself
Action =	what the document does
Audience =	document's *primary* audience
Object =	the key topic of interest to the audience
Outcome =	what the audience is able to do with the document's information

Warning: Do not omit any parts.

Error Trap: If you discover that you have two audiences, two objects and two outcomes, consider the possibility that you must write two documents or that you must write two distinct sections to a larger document. Treat each separately.

Discussion:

A well-written purpose statement helps you focus yourself as you research, organize, write, and revise. Craft your purpose statement by identifying each of the five parts:

- **Actor**—Identifying your document is not as simple as it might seem. Avoid writing staff studies and plans in memo or letter format.
- **Action**—Choose the action carefully because the action sets your tone. For example, a memo that *notifies* is more formal than a memo that *informs*.
- **Audience**—Write your document to your primary audience. Later you can edit to make the document more accessible to secondary and tertiary audiences if necessary.
- **Object**—Limit your document to one topic of discussion, else you'll ramble. Write about the topic of most interest to your audience.
- **Outcome**—Think through this key but often missed part. The outcome tells readers why they should read your document and sets their expectations. The outcome establishes a contract with your readers: you and your readers can decide if the document succeeds or fails to deliver on the contract.

Large documents like proposals have several actions addressed to different audiences, discussing different objects for different outcomes. Treat each major section of the large document separately, giving each its own purpose statement. For example, *This cost segment of the proposal details for the finance department the schedule and amounts of payments so finance professionals can budget accordingly.*

Example: Actors, Actions, Audiences, Objects, and Outcomes

Actor	Action	Audience	Object	Outcome
			procedures	
manual	instructs	client	annual budget	understand
tutorial	outlines	staff	design	perform tasks
proposal	explains	fellow experts	findings	make decision
letter	summarizes	lay persons	work plan	evaluate
memo	presents	supervisor	system	approve
study	notifies	likely buyers	job require-	keep for record
specification	demands	contract officer	ment	pursue further
bid	recommends	user	problems	plan accord-ingly

Suppose your analysis of purpose shows that you must write to a credit card company to get a bill adjusted. Your audience is the bank clerk who adjusts billing errors. You fill in the blanks and write the following purpose statement.

letter	notifies	bank clerk	billing error	adjust account
Actor	Action	Audience	Object	Outcome

"This letter notifies you (bank clerk understood) *about a billing error so you can adjust my account."*

See also: Purpose; Audience; Coherence—use front and back matter.

Step 3. Write a Purpose Statement

Exercise: Write a purpose statement after analyzing purpose and audience.
(Answer A-4)

Scenario 1: The building maintenance manager called to say they were scheduled to put sealant on the parking lot next week. Weather permitting, they plan to seal half the lot (the east side) on Monday, and the other half (the west side) on Thursday. The sealant dries in two days. They plan to paint lines on Saturday.

Therefore, only half the parking lot would be available Monday through Friday. The entire lot would be closed on Saturday and reopen the next Monday morning. The building manager asked you to inform your employees and ask for their cooperation. He also said that your lease gave the landlord the option of passing any towing fees to the company. He would contact the other fourteen tenants. Throughout the next week, parking space allocations would not apply and all parking would be first-come-first-served.

You have 320 employees in the building and you decide it would be most economical and accurate to give each employee a memo.

Actor	Action	Audience	Object	Outcome
		Ex: Employees		
		or		
		Employees who drive + park		

Scenario 2: You are building an E-mail system for your client. Your client asks you to submit a change proposal to expand the scope of your work to include an electronic bulletin board that will allow users to share information on technical topics.

Your staff believes they can add the bulletin board function for less than $80,000 for code and $18,000 for additional hardware, but the proposed new work extends the final system delivery by 120 days. You want the additional work. Write a purpose statement for the change proposal.

Actor	Action	Audience	Object	Outcome

Exercise: Evaluate these purpose statements. Are all five parts included? Do the five parts clearly and logically state the document's purpose? (Answer A-5)

1. This magazine article (*actor*) explains (*action*) to the general reader (*audience*) the advantages of recycling paper (*object*) so residents know how to cooperate in the county's recycling program (*outcome*).

2. This user guide informs new employees about procedures for using E-mail so they can send and receive E-mail.

3. This proposal explains our technical approach for the client users, and it presents our concept for the management plan, schedule, and costs for the financial department, and it justifies our bid in terms of time and materials, so you have the necessary information to evaluate our qualifications to successfully complete your project.

4. I thought I'd like to write down some thoughts and concerns for you in regard to the TDMS as we approach trials. We need to be constantly aware of several things, monitor their progress, and/or verify the fix.

5. This letter is a follow-up to our phone conversation today regarding the above captioned.

Technique 3.2	**Use the purpose statement to focus yourself and others.**

Tips: If you have a team writing the document, use the purpose statement to focus your team.

If your supervisor has an interest in your document, and you have any doubt that you and your supervisor see eye to eye, use the purpose statement to get agreement.

When writing deliverables, if you have any doubts that you and your client see eye to eye, use the purpose statement to get agreement.

Warning: Do not worry now about the exact wording of your purpose statement. Concentrate on getting agreement with your team, supervisor, and client. Edit later.

Example: You show the boss your draft purpose statement: *This memo informs employees about the company's 401K plan so you can decide whether to participate.* Your boss alters your purpose statement: *This memo explains step by step how to enroll yourself in the company's 401K plan so you can take advantage of this benefit.*

See also: Purpose; Audience; Coherence.

Discussion:

Recall that your purpose statement serves as a contract between you and your reader. If your reader rejects the premise of your contract, then your document is doomed. At present, you are only 5% into your writing assignment. You'll save time catching fatal flaws now rather than waiting for the final draft.

The purpose statement forces people to make firm decisions about the document. Consequently, the purpose statement uncovers fuzzy thinking and vague requirements. Despite the difficulty, press for concrete and specific guidance. Inevitably, you save time and effort.

For large documents such as proposals, your team members may come up with different purpose statements for their respective sections. That's OK. Sections of large documents often address different audiences with a different purpose. Just make sure each member of your team understands each other's purpose statement to avoid duplication and holes in the document.

Sometimes you are issued a purpose statement, perhaps a vague or inappropriate one. When questioning an assigned purpose statement, examine all five parts to ensure that the purpose statement is complete, and get agreement that the five parts work well together.

Exercise: Help B. Guirre devise a purpose statement for this memo. How does a purpose statement focus B. Guirre? (Answer A-6)

Memorandum

To: K. Lemnick, ACME Temps

From: B. Guirre, Project Manager, XYZ, Inc.

Subject: Employment of Temp (Ida Smith)

I contacted ACME Temp during the month of August seeking someone who would be knowledgeable about both basic graphics software and WordPerfect 5.1. ACME Temp sent temporary assistant Ida Smith who was, in fact, a fine match for our project.

Ida was well aware that September 19th was the deadline for our project. September 19th also happened to be the same day she planned to go out of town on vacation. To ensure that the deadline would be met, and to accommodate Ida's schedule, we agreed to special working conditions for the weekend of September 15th and 16th. For this weekend only, she was provided with a building pass, my home phone number, and instructions to contact me if she had any problems. Therefore, she could complete her work and leave for vacation as planned.

On Monday, I received a note from Ida, by means of courier, explaining that she had not completed the project as agreed, because she had left town early for her vacation. By failing to contact me earlier, and by subsequently leaving town, she left me with an unfinished project and no time to properly train someone to take her place.

Ida was our first contact with ACME Temps for a major project, and <u>we did not</u> have a good experience. Please follow up on this situation as soon as possible.

Actor	Action	Audience	Object	Outcome

Step 3. Write a Purpose Statement

Final exercise: Analyze purpose and audience, then write a purpose statement for the system design. (Answer A-6)

Scenario:
Your team won the contract to build a computerized "Fulfillment and Invoicing" system for Pinnacle Trail, Ltd., a fast-growing catalogue sales company specializing in hiking and camping equipment. In your technical approach, you bid that you would acquire a UNIX-based software package that you would then customize to meet Pinnacle's special needs.

The next deliverable for the contract is a detailed system design from which your technical staff will write code to modify the UNIX-based package. The Pinnacle Information Systems (IS) manager, who represents the Pinnacle users, will approve your detailed system design. The Pinnacle V.P. of Sales and Marketing, who is nervous about *"all this hi-tech stuff,"* also wants to see the system design so he can understand what's going to happen to his slow but effective low-tech procedures.

Analyze purpose.

- Your purpose for writing:

- Reader's purpose for reading:

- Purpose of the work:

Analyze audience. Profile each audience using a checklist.

Audience	?	?	?
Priority	Primary	Secondary	Tertiary
Reading as			
Profession			
Knowledge of subject			
Friendly, hostile, neutral			
Considers you expert			
Wants background			
Wants theory			
Wants examples, pictures			

Write a purpose statement.

Actor	Action	Audience	Object	Outcome

Step 4. Gather Information

3. Write Purpose Statement

4. Gather Information

Use purpose statement as guide

Ask 5 Ws & H

Having defined your document with a purpose statement, gather information using two techniques:

4.1 Use your purpose statement as a guide as you write down ideas and assemble facts.

4.2 Ask *Who, What, Where, When, Why,* and *How* to ensure you get all the details.

Discussion:

Most technical experts became technical experts because they like research. Consequently, they often dive into the research before defining the purpose of the document. Then they record the fruits of their boundless research in the document.

From the audience's point of view, unfocused research wastes their time as well as the writer's. *"I just wanted to know what time it is, and you told me how to build a clock."*

Technique 4.1	**Use your purpose statement as a guide as you write down ideas and assemble facts.**

Tips:	Use your purpose statement and your audience profile to focus information gathering. For example, if you know your reader doesn't want theory, then you need not gather details about theory for your document.
	Write down ideas, facts, and details in complete short sentences using short words.
Warning:	Do not try to organize ideas yet—you sort and order ideas later.
	Do not write in paragraphs yet—you write the draft later.
Example:	Consider gathering information to support the following purpose statement.

This manual *(actor)* describes *(action)* to new owners of the Fastcut Lawnmower *(audience)* proper maintenance and safe operation of your new lawnmower *(object)* so you can get more out of your machine *(outcome)*.

1. *Actor*—manual—suggests a brief *how to* document. Gather information on procedures.
2. *Action*—describe—suggests that we need not explain the mechanics or attempt to motivate the new owner. Gather specific details that describe the lawnmower.
3. *Audience*—new owners—indicates a operator who has no interest in the theory of small four-cycle engines. Gather only practical information.
4. *Object*—maintenance and operation of the new machine—limits the discussion. You need not discuss the history of lawn mowers, for example.
5. *Outcome*—can get more—suggests details must be complete and anticipate readers' needs as they try to operate and maintain their machine.

See also:	Audience; Purpose statement.

Discussion:
Expect to gather more information than you use in your draft. Later when you organize your ideas, you select the most useful details and discard the superfluous.

Exercise: Describe how the following purpose statements focus information gathering. (Answer A-7)

Scenario: You are a real estate manager and you must write two letters that describe the same empty office space to different audiences for different purposes. You write a purpose statement for each letter.

1. This bid request specifies to ACME Carpet Co. the dimensions of eight vacant offices at 123 Maple Avenue so your carpet company can bid on installing wall-to-wall carpeting.

    Actor—
    Action—
    Audience—
    Object—
    Outcome—

2. This brochure outlines for prospective renters the features and benefits of our vacant offices at 123 Maple Avenue so prospective renters can decide whether to schedule an appointment to view the office space.

    Actor—
    Action—
    Audience—
    Object—
    Outcome—

Technique 4.2	**Ask *Who, What, Where, When, Why,* and *How* to ensure you get all the details.**

Tips:	Ask *who, what, where, when, why,* and *how* (5 Ws and H) questions from the reader's point of view.
	Write down all questions that come to mind. Expect some redundancy.
	Gather information to respond to your questions.
Warning:	Don't waste time answering questions that your purpose statement shows are irrelevant.
Example:	The following purpose statement may generate these questions:

This letter describes for prospective renters the vacant offices at 123 Maple Avenue so renters may appreciate the many fine features and benefits that come with the office space.

Who are the prospective renters and who are we?

What features and benefits interest renters?

Where is 123 Maple Ave? *Where* can the prospect contact us?

When does the offer expire? *When* can renters move in?

Why must renters act now?

How can a prospective renter take advantage of this offer?

See also:	Audience; Purpose statement.

Discussion:
Using the *who, what, where, when, why,* and *how* questions helps you gather information readers need, not just what you want to tell them. These questions help you think of details you might otherwise forget.

Exercise: Ask *who, what, where, when, why,* and *how* questions to generate details for documents with these purpose statements. (Answer A-7)

1. This tech-report describes the repairs and tests we conducted on your PDQ Laserprinter to stop the intermittent errors you reported, so you know what was covered by warranty and what service you must pay for.

2. This guide provides you, the new owner of a Megawheel plastic tricycle, simple step-by-step assembly instructions so you can quickly put your little tyke on his Megawheel trike.

3. This letter notifies LD Cellular Phone Co. billing office about $12,456.56 of unauthorized calls charged to our account so you can change our phone access code, investigate the fraudulent calls, and remove the charges from our account.

Final exercise: Use your purpose statement and the questions *who, what, where, when, why*, and *how* (the 5 Ws and H) to focus information gathering.
(Answer A-8)

Scenario: You are a regional sales manager for a rapidly expanding chain of hardware stores. You have 20 stores in your region, each with a manager. You send a monthly report telling store managers when to expect new products, sales promotions, and price changes. You also tell them how to set up displays and stock shelves, and alert them to hot trends in hardware merchandising.

Your purpose statement reads: This monthly marketing report informs you of new products, prices, and promotional efforts so you can prepare your store to take maximum advantage of sales opportunities.

1. How does the purpose statement focus your information gathering? Also, give one example of data you don't need to gather.

      Actor—
      Action—
      Audience—
      Object—
      Outcome—

2. Using the 5 Ws and H, what information do you gather?

Step 5. Write Sentence Outline

```
┌─────────────────┐
│   4. Gather     │
│  Information    │
└─────────────────┘
         │
         ▼
┌─────────────────┐
│   5. Write      │
│   Sentence      │
│   Outline       │
└─────────────────┘
┌─────────────────┐
│     Write       │
│   assertions    │
└─────────────────┘
┌─────────────────┐
│    Evaluate     │
│   assertions    │
└─────────────────┘
┌─────────────────┐
│  Put assertions │
│    in order     │
└─────────────────┘
```

After gathering information, use a sentence outline to get your ideas down and put them in effective order.

Use three techniques to write a sentence outline:

5.1 Write your ideas as assertions using short words in short sentences.

5.2 Evaluate assertions against your purpose statement.

5.3 Put your assertions in effective order.

Discussion:
Studies show that people can concentrate on organization or details, but not both at the same time. It's true: we can see the forest *or* the trees, but not both. As a writer, first determine your organization, so secondly you can concentrate on details while you compose your draft.

Technique 5.1   **Write your ideas as assertions using short words in short sentences.**

Tips: Use your gathered information as the source for your assertions.

Write each assertion as a short sentence (3 to 10 words).

Use short words that clearly—bluntly—assert your idea.

Limit your short sentences to expressing assertions. Save the details for later.

Warning: *Do not* write a topic outline. A topic outline tells you only vaguely what you're going to write about, not the point you want to make about the topic. Topic outlines result in omitted thoughts and unnecessarily repeated thoughts.

Do not write a stream-of-consciousness draft instead of a sentence outline. A stream-of-consciousness draft often reads like a story—how events or thoughts occurred to the author. Irrelevant and out-of-sequence thoughts result.

Example:

Topic outline: Schedule

Stream of consciousness: When we first learned of your accelerated launch schedule, we naturally gave the matter serious consideration. . .

Sentence outline: Our team can meet your launch date.

See also: Gather information.

Discussion:

After gathering information, you must select, evaluate, and organize your ideas so your reader can understand and use your information.

Write your ideas as assertions. An assertion differs from a topic, fact, or detail because you can argue with an assertion, and you can't argue with a topic, fact, or detail. Your sentence outline makes assertions that you and others can evaluate. A sentence outline built of assertions helps you get to the point.

Write your assertions with short words in 3-to-10 word sentences. Some writers put assertions on 3x5 cards, others use stick-on notes—just use anything you can later sort. Concentrate on getting your ideas down as simply as possible. You can edit the word choice later.

Exercise: Using gathered information as your source, write assertions for a sentence outline. Use short words in short sentences. (Answer A-9)

Scenario:
Susan Walters recently ran a successful pilot program for job sharing for BAP Industries. Management asks her to write an article for the company magazine. Susan writes a purpose statement and gathers information. Help Susan begin her sentence outline by writing short assertions based on each group of information below. These assertions later become part of Susan's sentence outline.

Purpose statement: This article describes how BAP Industries recently implemented a successful pilot program for job sharing, so you can see if job sharing meets your or your department's needs.

Information gathered:

1. *Who* applied for job sharing? Twenty BAP employees and ten non-employees applied for the job sharing pilot program. Applicants varied from a part-time attorney to a full-time senior manager. Most applicants were administration and human resource department professionals with a range of experience.
Assertion: Many competent applicants with a wide range of experience applied.

2. *What* are the advantages and disadvantages of job sharing? The corporation retains good employees who know the job and the organization. Job sharing entails little extra cost. Job sharing increases employee morale. Management can lose control. Customers can become confused. Job sharers can be incompatible.
Assertion:

3. *Where* does job sharing occur? The percentage of U.S. companies that job share by region: West–8%; Mid-west–13%; South–6%; Northeast–7%; Great Lakes–16%.
Assertion:

4. *When* do job sharers work? Each partner works 20 hours a week and earns 1/2 company benefits. Each partner works a $2\frac{1}{2}$ day schedule. Partners work 2 hours together mid-week to ensure continuity.
Assertion:

5. *Why* did applicants want to job share? Some wanted time to pursue personal interests like education. Some working parents wanted more family time.
Assertion:

6. *How* did job sharers communicate the new working arrangement to others? They published internal memos introducing the change. They sent a letter to vendors and clients. They reminded contacts of the job sharing arrangement when using voice mail, telephone, or written communication.
Assertion:

Technique 5.2	**Evaluate your assertions against your purpose statement.**

Tips:	Examine your collection of one-sentence assertions written on separate cards.

1. Eliminate obvious irrelevancies. If the assertion does not support your purpose statement, the assertion is probably irrelevant.
2. Eliminate redundancies. If two assertions say basically the same thing, get rid of one.

Warning:	Do not expect to get the assertions down exactly right the first time. Writing the assertions down gives you a chance to evaluate them. Expect to throw some away and combine others.

Error Trap:	If too many assertions don't support your purpose statement, consider the possibility that your purpose statement is flawed.

Example:

Purpose Statement: This memo announces to all employees changes in the cafeteria's hours so they can plan their lunch breaks.

Assertions (some deleted as irrelevant or redundant)
~~All employees have the option of eating in the cafeteria.~~ *irrelevant*
~~Some employees bring lunch and eat outside instead.~~ *irrelevant*
We shortened the cafeteria's hours to 12:00 P.M.–2:00 P.M.
The cafeteria is understaffed.
~~The cafeteria had to lay off workers.~~ *redundant*
We expanded the cafeteria's dining area.
We regret any inconvenience caused by the shortened hours.

See also:	Purpose statement.

Discussion:
Use your purpose statement to determine which assertions belong in your discussion:
1. Actor—Different documents and formats, such as proposals and manuals, demand certain assertions.
2. Action—Do your assertions support the document's action and tone?
3. Audience—Does the audience accept your authority, or must you back up all assertions?
4. Object—Do your assertions develop the key topic, or do they raise irrelevant side issues?
5. Outcome—Do your assertions move the reader to the intended outcome? If not, your assertions lack relevance.

Exercise: Evaluate the following assertions against the purpose statement. Eliminate irrelevancies and redundancies. Remember to evaluate assertions in this order:
    1. Find and eliminate irrelevancies based on your purpose statement.
    2. Find and eliminate redundancies by inspecting remaining assertions.
Tip: Concentrate on costs and benefits so the audience can decide whether to invest. (Answer A-10)

Purpose Statement: This fact sheet highlights for senior managers the costs and benefits of our proposed automated data security system so you can decide whether to invest in the new system.

Assertions:

Ninety percent of Fortune 500 companies use data security systems.

Our professionals keep a lot of valuable information on our computers.

We rely on computers now more than ever to be profitable.

Losing data can severely reduce profits.

The proposed system uses three optical drives tied to our local area network.

The next-best alternative used old technology, a 16 BPI tape drive.

The 3 optical drives, 10 cartridges, optic fiber cables, and software cost $6,490.

Causes for data loss range from employee error to natural disaster.

Industry surveys provide statistics on industry-wide information loss.

The proposed system limits data loss to a worst case 24 hours.

Last year, lost data cost us more than 3,500 direct labor hours at $40 per hour.

Our costs for losing data exceeded industry averages.

Expect our proposed data security system to reduce data losses by 85%.

Lost data results in misplaced or late orders, hence angry customers.

Technique 5.3	**Put your assertions in effective order.**

Tip 1: Organize with your primary audience's needs in mind. Recall your audience profile.

1. Determine main assertions and place supporting assertions under the main assertions.

2. Find and correct omissions and subtle irrelevancies.

- If a main assertion lacks subpoints, you must either add the omitted subpoints, or delete the main assertion as irrelevant.

- If subpoints lack a main assertion, you must either add the omitted main assertion, or delete the subpoints as irrelevant.

3. Order your grouped assertions the way your reader wants to see them.

Warning: Do not start writing your draft until the sequence of ideas makes sense. You can't edit logical sequence into a document after writing the draft.

Example: Consider the assertions in Tip 1 above: Note the main and supporting assertions, and note that assertions are ordered as the reader will use them.

See also: Purpose; Audience; Purpose statement.

Discussion:

Effective order meets the reader's needs. Recall that we can write to only one audience at a time. Note how audiences have different, even incompatible needs.

Weigh the audience's needs against your own reasons for writing as you order your ideas.

Sentence outlining provides these four benefits:

1. When you complete your sentence outline, you can show it to to your team, supervisor, or client to get feedback. Together you find omissions, irrelevancies, redundancies, and out-of-sequence assertions before you invest further effort in a draft. With a topic outline, people have to guess what you intend to say. With a stream-of-consciousness draft, people either can't find your ideas, can't follow them, or won't take the time.

2. Know that the assertions in your outline become the key sentences for your draft. Having thought through your assertions, you can concentrate on the supporting details when writing the draft.

3. Your document is easier to read because your key assertions are clearly expressed in short sentences with short words at the beginning of paragraphs.

4. Your sentence outline can serve as a summary of your document later, because it includes all your key assertions in sequence.

Exercise: Determine the main assertions and supporting assertions. Group supporting assertions under main assertions. Put groups in order. (Answer A-11)

Scenario: Your New Jersey company plans to build a can manufacturing plant somewhere in the Southeast. You must write a staff study to the company president showing the pros and cons of building the plant in Tuscaloosa, AL.

Purpose statement: This staff study details to you (the president) the pros and cons of building a can manufacturing plant in Tuscaloosa, AL, to help you decide where to locate the new plant.

1. The Black Warrior River-Tombigbee Waterway offers easy access to the Port of Mobile.

2. At present, Tuscaloosa's business climate meets our needs.

3. Tuscaloosa provides few big-city amenities.

4. Tuscaloosa has no big-city costs.

5. The University of Alabama offers some cultural opportunities.

6. The Tuscaloosa Airport requires connections through Atlanta.

7. The local area costs are low for our business.

8. Personal income, property, and sales taxes are low.

9. Recent layoffs in local chemical, rubber, paper, and iron factories make an abundance of cheap, skilled labor.

10. Tuscaloosa has many recreational facilities: lakes and parks.

11. Tuscaloosa has mild winters and hot summers.

12. State and local governments offer five-year tax incentives to move into Alabama to prop up the stagnant economy.

13. Local public schools are rated below average, but improving.

14. Tuscaloosa's transportation is geared for manufacturing companies.

15. Tuscaloosa serves as a minor hub for both highway and rail traffic.

16. Tuscaloosa presents a major change in living environment for our New Jersey transplants.

Tip 2:   Use natural patterns of thought and standard formats to order your assertions, because they are easy for you and your reader to follow.

Warning: Do not present information chronologically unless you have a good reasons for doing so. Avoid telling a story—explaining the situation as it happened to you—instead of ordering the information to best serve the reader. Chronological presentations sound like our old grade school essays: "How I spent my summer vacation."

Example: Natural patterns for presenting assertions and details include these:

Chronological	*(First this, then that. . .like a history)*
Ordinal	*(Step 1, step 2. . .like a cook book)*
Priority	*(Greatest to least, or least to greatest)*
Functional	*(How things work. . .like describing a computer)*
Topical	*(Attribute 1, attribute 2. . .like a list)*
Spatial	*(Top down, or bottom up. . .like describing a house)*
Examples	*(Generalization illustrated with supporting details)*
Cause-Effect	*(Also problem-cause-solution or effect-cause)*
Compare	*(Also contrast or compare-contrast)*

See also: Purpose statement; Revision—content and organization tests.

Discussion:

Readers and writers are accustomed to certain natural patterns used to present information. These patterns reflect the ways our minds work; they are practically second nature to us. Writers have also developed standard formats to respond to recurring business writing situations.

Use natural patterns of thought within standard formats whenever possible. Someone else has thought through your writing situation, saving you time and effort. Also, standard formats help your readers, because they know what to expect in the document and how to use it. For example, a status report's standard format might use an overall topical pattern (actions completed, actions ongoing, actions planned) with a priority pattern used within each topic.

Different organizations have their own standard formats for similar documents. For example, the Army staff study format—problem, facts, alternatives, solution, discussion—differs from most business staff study formats—problem, solution, discussion, facts.

Exercise: Identify the natural patterns used to order each group of assertions and details. (Answer A-12)

Group 1

1. The auto-transmission is a driveline component between the engine and the drive axle.

2. The auto-transmission is a major component of the drive train.

3. The transmission converts engine horsepower to torque delivered through the drive shaft to the axle.

4. Unlike standard transmissions, the automatic is placed in gear and remains in that position until a change is required in the vehicle's mode of travel.

5. The auto-transmission consists of portioning valves, metering bodies, clutch disc packs, and multiple pumps.

6. Auto-transmissions range from large and bulky to small and lightweight.

Group 2

1. Auto-transmissions require very little maintenance.

2. Maintenance would include a yearly service, which consists of removing the transmission pan, replacing the filter, gasket, and transmission fluid.

3. During maintenance, a general inspection of the valve bodies should be conducted when the transmission pan is removed.

4. The transmission pan is inspected for any foreign substance that would indicate there may be internal problems with the component.

5. Any foreign material found in the pan would be an early indicator of future problems with the transmission.

Group 3

1. The first auto-transmission was the *powerglide*. This two-speed transmission was used in the early years to provide more luxury to the automobile.

2. Vehicles increased horsepower and greater torque, which called for a stronger component within the drive train.

3. The turbomatic was developed and designed as a three-speed auto-transmission providing a greater range of torque.

4. Throughout the years this automatic transmission proved itself, through testing and use, to be one of the finest components ever designed.

5. Today the turbo transmission is used in all makes and models of cars and trucks. This transmission is used today in both front-wheel and rear-wheel drive trains.

Final exercise 1: Write a sentence outline. (Answer A-12)

Scenario: You are the Project Manager for a $3 million effort to upgrade the Universe Bank Credit Card transaction and billing system. When completed, your new system will increase the transaction throughput from 7,500 to over 200,000 transactions per hour. The greater throughput reduces transaction costs, which could save Universe Bank up to $120,000 each month. Senior bank management asked you to accelerate development to install the system no later than November 20, in time for the heavy Christmas volume.

Your senior engineer tells you that the project is on schedule, more or less, with some minor interface problems that slow the system down. Your technicians cannot accelerate delivery unless they take short-cuts on bench-testing the software. Even then, they expect the new Bank Card system will need a 45-day shakedown to work out problems.

You think the Christmas season would be the worst possible time to install, then "debug" a new, therefore, unstable Bank Card system. Your engineer estimates that the new system might (worst case) crash often enough to be off line as much as three hours per day. At 7,500 transactions per hour, averaging $0.20 per transaction, Universe Bank could lose as much as $1.3 million in revenue, not to mention the loss of good will with their Bank Card merchants, who live or die on their Christmas retail trade.

You decide to write a letter to the Universe Bank Informations System (IS) manager. Analyze purpose and audience, write a purpose statement, then write a sentence outline.

1. Analyze purpose: your purpose for writing; reader's purpose for reading; purpose of the work.

2. Analyze audience: primary, secondary, and tertiary. Profile each audience using a checklist.

3. Write purpose statement.

Actor	Action	Audience	Object	Outcome

4. Write a sentence outline. Write assertions, eliminate irrelevancies and redundancies, check for omissions, then put assertions in effective order.

Final exercise 2: You get tasked to send out the annual holiday season letter to vendors explaining your company's policy on gift-giving. You dredge up last year's boilerplate letter. Use your sentence outlining skills to disassemble and then reassemble this poorly organized letter. (Answer A-13)

> First, write each assertion you find in the letter as a short sentence.
> Second, analyze purpose and audience.
> Third, write a purpose statement for the letter. (Don't try to make your purpose statement accommodate all the original assertions.)
> Fourth, using your purpose statement, evaluate and order assertions in a sentence outline.

December 1, 1994

To: Our Vendors

With the approach of the Holiday Season and the close of 1994, it is appropriate that we should write you and express our appreciation for the past performance of your company. Your attention to our purchasing contractual terms and conditions in the areas of product quality, reliability, and timely response to our requests is very much appreciated by all of us at BAP.

It is not at all an uncommon practice for suppliers to express their appreciation to personnel in customer organizations by remembering them with a holiday gift. We have concluded, from both an ethical and a strictly business point of view, that anything more substantial than an advertising novelty, whose value is minimal, or a card expressing appreciation would be inappropriate. We hope that you will reach this same conclusion as one of our suppliers.

Our policy with respect to gifts is not intended to imply improper behavior on the part of our suppliers or employees; but only to emphasize our wish that suppliers concentrate their expenditures on our behalf entirely on improving the quality and cost of products and services they sell us.

Thank you for your cooperation in this and other matters within the scope of our business relationship, and our very best wishes to you for the Holiday Season and the New Year.

Sincerely,

Bob Cratchit
Accounting

Writing Phase

The Writing Phase relies on your composing skills as you develop your sentence outline into paragraphs, using your gathered information as supporting detail.

Step 6   Write the Draft

Discussion:

During the Writing Phase, you use your composing skills—one of three distinct skill sets. Unfortunately, many writers try to analyze, compose, and edit while writing the draft. A writer's speed is determined in large measure by how the writer uses the writing system and separates skill sets.

Most writers fall into one of three categories:

- **Sprinters** write the draft without regard for word choice, grammar, punctuation, or mechanics, and they do not stop to analyze content.

- **Plodders** analyze an assertion, write the corresponding paragraph, then edit the paragraph. Plodders use their analytical, composition, and editing skills as they move from paragraph to paragraph.

- **Bleeders** analyze, write, and edit each sentence, making sure they have a Pulitzer Prize-winning sentence before moving on to the next.

Plodders and bleeders waste time and energy shuffling between skill sets. And after finishing the draft, even the bleeder must revise, edit, check for correctness, and proofread.

If you're a plodder or a bleeder, resolve to become a sprinter. You can increase your writing speed by as much as 50%.

6. Write the Draft

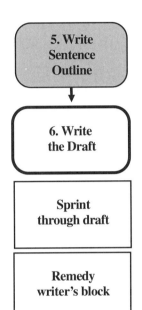

With your sentence outline completed, use two techniques to write a clear draft:

6.1 Sprint through the draft.

6.2 Identify sources of writer's block and apply remedies.

Discussion:

When writing your draft, get your information down as quickly as possible so you don't lose your train of thought.

You can write your draft quickly because you have a good sense of purpose and audience. With your purpose statement, you focus your writing task to a manageable project. You have pertinent information and a complete sentence outline of relevant and supported assertions. You are ready to write your draft.

In your draft, you develop assertions into paragraphs using your gathered information as supporting data.

Technique 6.1	**Sprint through your draft.**

Tips: Practice these seven tips to sprint through your draft:

1. Put yourself in a good environment—quiet, well-lit, and physically comfortable.
2. Get information down in the first words that come to mind. Use short words and short sentences. Write with total disregard to word choice, grammar, punctuation, and mechanics.
3. Write down all the details that you can think of to support each assertion in your sentence outline. You can cut information later when you have the whole picture.
4. Use personal shorthand to get information down as quickly as possible. Don't stop to grope for that perfect phrase, word, or detail. Use open and close brackets to mark the "hole." Come back and fill it later.
5. Take a sectional approach, writing the easier sections first. Usually, you write the body first, then the conclusion and lastly the introduction, because the body is easiest to write and introduction the hardest.
6. Take short (ten-minute) breaks every hour.
7. Use tools to make the job easier. Use a keyboard or paper, whichever works best for you.

Warning: Don't permit interruptions—you lose thoughts and time.

Don't give into temptation to edit before you finish the draft. Have confidence in the writing system in which you methodically find and correct errors.

Don't break when you're stymied. You find yourself trying to think through problems when you should be resting your tired brain.

See also: Audience; Purpose statement; Gather information; Sentence outline.

Discussion:

Do a thorough pre-write so you can stay focused when writing your draft. Your document can succeed at one purpose. You can write to only one audience at a time. Later, when editing for coherence, you use techniques to make your document more accessible for secondary and tertiary audiences.

Technique 6.2   Identify the source of your writer's block and apply remedy.

Tips: Recognize these five sources of writer's block and apply the corresponding remedies.

1. *Lack of well-defined purpose.* You find yourself asking, "Why am I writing this?" You need to go back to the purpose statement.

2. *Poor knowledge of audience.* You find yourself trying to write to multiple audiences at once, an impossible task. You need to analyze audience, write a purpose statement, then organize your content for your primary audience.

3. *You lack information or have nothing to say.* You don't have a sentence outline made of assertions. If you lack enough information for a sentence outline, you need to Gather Information, and if you can't determine what information to gather, you must backtrack to purpose statement.

4. *You lack confidence.* You don't know why you're writing, or to whom, or how they will use the document, therefore you don't know what to say. Go back to analyzing purpose, and work through the purpose statement to a sentence outline.

5. *Fatigue.* You're pleased with your pre-writing, but you're too tired to pick up a pencil. Take a break.

Warning: Do not waste time trying to write through a writer's block. If you're tired, you write poorly. If you need to gather information or define purpose and audience, you write the wrong message to the wrong audience for the wrong reasons.

See also: Audience; Purpose statement; Gather information; Sentence outline.

Discussion:

Writer's block is getting stuck—staring at the paper or screen—while writing the draft.

Writer's block serves as an *error trap* to uncover problems in the pre-writing phase. Writers who skip analysis of purpose and audience, skip writing a purpose statement, or skip the sentence outline suffer disproportionally more writer's block. In fact, the problem is really analyst's block.

To help reduce blocks caused by writer's fatigue, use the writing process to allocate your time. If you have only two days to write a proposal, you know you must have your sentence outline done halfway through day one, and so on. If you try to cram a four-day writing project into two days, you get tired, and quality suffers.

Post-writing Phase

The Post-writing Phase includes all the work you do *after* writing the draft. Use your editing skills to evaluate and revise content and organization; edit for coherence, clarity, economy; check for readability and correctness; and finally, proofread.

Step 7   **Revise Content and Organization**

Step 8   **Edit for Coherence**

Step 9   **Edit for Clarity**

Step 10   **Edit for Economy**

Step 11   **Check Readability**

Step 12   **Check Correctness**

Step 13   **Proofread**

Discussion:

You are the best person to revise and edit your draft. Who else is as qualified to

- evaluate content and organization
- emphasize the logical flow and highlight important points
- settle controversies about clarity
- ruthlessly cut unnecessary words
- simplify the vocabulary and cut sentence length
- correct departures from standards
- proofread the final product

So, who's the best person to edit any boilerplate that you might have in your document? The author of the boilerplate, of course. But the boilerplate author disappeared many documents ago, so you must use extra care to edit any boilerplate you use and make it your own.

Step 7. Revise Content and Organization

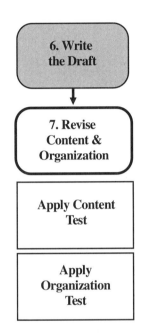

After writing the draft, use two techniques to revise content and organization:

7.1 Apply the three-part Content Test.

7.2 Apply the three-part Organization Test.

Discussion:

Revision means re-seeing. Take a break after writing the draft to refresh and distance yourself from your writing. Work with a double-spaced paper copy of the draft so you have plenty of room to annotate. Put yourself in the reader's place:

    What questions does the reader need answered?

    What organization suits the reader's needs?

The Content and Organization Tests serve as major error traps. Note that with each part of each test, we recommend corrective action. When you ask others to review your draft, ask them to use the Content and Organization Tests so you get precise, actionable feedback.

Then revise your draft before you edit for style. Editing a document that has content and organization problems wastes time and effort.

Technique 7.1	**Apply the three-part Content Test.**

Tips: Apply the Content Test by answering three questions:

1. What is the topic?
   Does the document reveal its topic within the first two sentences? Look for key words that name the topic. If not, use a purpose statement in the beginning to focus the reader.

2. So what?
   For each assertion and detail in the document, ask "*so what?*" Do readers care? If not, cut the irrelevant material or adjust the purpose statement to show the relevance.

3. How supported?
   For each assertion that passes the "*so what?*" part of the test, ask "*how supported?*" Do you lack supporting detail to make the assertion believable? Are your details unclear to your audience? If so, you may need to gather more information or cut the unsupported assertions.

Warning: Do not confuse the topic of interest to you with the document's key topic. For example, when you write a letter of application, *getting the job* most interests you, but *ability to meet job requirements* is the document's key topic.

Example: The poor version fails all three parts of the Content Test.

Poor: The building manager called to say that our parking garage has leaks. Therefore, he contracted a paving company to fix the leaks next Monday. Therefore, make other plans to park.

Good: This memo informs you that you can't park in our garage Monday. Please make other arrangements. Remove your vehicle from the garage by 8:00 P.M. Sunday. After needed maintenance, the garage re-opens 6:00 A.M. Tuesday.

See also: Audience; Purpose statement; Gather information.

Discussion:

The Content Test anticipates three questions asked by critical readers.

You can effectively run the Content Test with one quick read of the document.

The Content Test will catch any logical fallacies in your argument. However, the Content Test does not catch a lie, so you may wish to add a fourth query: *"Is the assertion true?"*

To pass the Content Test every time, write a good purpose statement during the pre-writing phase and put it at the beginning of your document.

Exercise: Apply the Content Test to this letter. Suggest ways to improve content. (Answer A-14)

    Question 1: What is the topic? Do key words in the first two sentences make the topic clear? If not, suggest a purpose statement.

    Question 2: So what? Delete assertions that fail the *so what?* part of the test.

    Question 3: How supported? Suggest supporting details where an assertion needs to *specify how*.

BAP Industries, Inc.
1 Liberty Plaza, McLean, VA 22036
(703) 555-1234

April 4, 1993

Washington National Bank
Customer Service
P. O. Box 12345
Washington, DC 20005

Dear Customer Service:

Enclosed please find a copy of our recent VISA bill. The transaction that I have circled is the transaction we are disputing. The date of the transaction was August 5th and the date it was posted was August 8th. The company was Greenspan & Co., Inc. in Tuscaloosa, Alabama. I have never heard of this company before. The amount charged was $520.97. That is why I am questioning this transaction. Is there any more information that you can send us regarding this transaction?

Our accounts payable department and I have no idea where this bill has come from. We have a purchase order system and I would have known if I had approved any request involving the Greenspan company. I didn't. It is possible that our number was mistakenly used or punched into the computer. I personally know how easy it is to transpose numbers. Your help in any way would be very appreciated in this matter.

Sincerely,

Dennis Smith
enc: VISA bill

Technique 7.2	**Apply the three-part Organization Test.**

Tips: Ensure that your message organizes the content in a way that helps the reader understand and respond. Apply the Organization Test by answering three questions:

1. Does the document read like a data dump?
    Does the document wander randomly from idea to idea? If so, you probably skipped sentence outlining and relied on stream-of-consciousness writing.

2. Does the document read like a story?
    Does the document present events, research, or ideas as they occurred to you? If so, you probably need to re-order your assertions in a way useful to the reader.

3. Is the document filled with *I, me, mine*?
    If so, you may be venting. You wrote your draft from your point of view—how the circumstances affect you, rather than how the circumstances affect the reader.

    If your document fails one of these organization tests, go back to sentence outlining. You can take the pieces of your document that passed the Content Test and re-sort them as major and supporting assertions.

Warning: Do not try to edit your way out of failed organization: You won't succeed.

Example:

Poor: I just received your notice of overpayment and a refund check. I wrote in April asking you to apply the money to my next bill. Obviously you didn't. I didn't want a refund. I can be reached at my business number if you have questions.

Good: Please apply this returned refund check to my next bill as requested in my April letter.

See also: Sentence outline.

Discussion:

You can effectively run the Organization Test with one quick read of the document.

If you used a sentence outline, you should pass the Organization Test.

Conversely, if you fail the Organization Test, you must return to the sentence outline step. You can not patch or edit your way out of an organization problem. You must disassemble, then reassemble your document. Break your document into assertions, one per card, then follow sentence outlining instructions to reassemble.

Exercise: Joan Walters asks you to critique her letter. Apply the Organization Test and give her specific advice how to improve her letter. (Answer A-14)

Question 1: Does the document read like a data dump?
Question 2: Does the document read like a story?
Question 3: Is the document filled with *I, me,* and *mine*?

Mr. Traver Stark
Vice President Government Sales
Aero Space, Inc.
1414 Vermont Ave.
Washington, DC 20016

Re: Internal Job Availability Notice B-22
     Marketing Analyst/Proposal Specialist

Dear Mr. Stark:

I currently fill the position of an Administrative Assistant for the Government Business Development (GBD) Department at Aero Space. It is my desire to seek a position that would allow me to fully utilize my communication and organizational skills.

I was very excited to learn that a position has opened in the Marketing Department for Foreign Sales. This position appears to be tailored specifically to my career goals at this time. It would afford me the opportunity to apply my knowledge of proposal preparation, and at the same time exercise my ability to coordinate projects and follow them through to completion.

As my resume indicates, my qualifications show that I am well experienced in this field. I have five years experience in proposal preparation. In addition, my exposure to foreign languages, my desire to travel, and my experience in U.S. Government sales have well prepared me for my next career move.

I look forward to meeting you to discuss my potential for securing this position.

Sincerely,

Joan Walters
enc: resume

Final exercise: Apply the Content and Organization Tests. Recommend ways to improve the letter. Work quickly—you need not spend much time running the tests. (Answer A-15)

TechtopIndustries, Inc.
1 Liberty Plaza, McLean, VA 22036
(703) 555-1234

December 12, 1994

General Warranty Ltd.
Customer Service
P. O. Box 12345
Racine, WI   40105

Dear Customer Service Department:

On December 21 of last year, we asked for reimbursement for an alternator repair to our new company car, which we purchased through a dealer who sold us your extended warranty. (See attached letter.) Now as a result of the defective alternator, the company car's battery died, and had to be replaced.

When the service center replaced my alternator, they had trouble recharging the battery, but they suggested I keep the old battery for awhile to see if it would hold a charge over a period of time. As it turned out, two weeks after the alternator was replaced, the battery died. Although it's December, the weather has been unseasonably mild; so I guess cold weather did not kill the battery. The battery died as a direct result of the previous defective alternator running it down over a period of time.

I realize that the battery is not specifically covered under our warranty, but because it died as a direct result of the mal-functioning alternator, I think it's only fair that you replace the battery. It was only 17 months old. We shouldn't have had alternator or battery problems in the first place. Please call me if you have any problems with this.

Sincerely,

Paul Jackson

Step 8. Edit for Coherence

7. Revise Content & Organization

↓

8. Edit for Coherence

Use key words

Group ideas

Preview with introduction

Use front & back matter

Apply visual devices

After you've checked content and organization, edit for coherence. Give verbal and visual cues that help your reader

- skim your document
- follow your discussion
- refer back to your document

Use five techniques to edit for coherence:

8.1 Use key words in titles, subheads, and throughout the document.

topic sentence lists

8.2 Group your ideas in paragraphs and vertical lists.

TOC

8.3 Preview your document or major sections within your document with an introduction.

8.4 Use front and back matter to help secondary and tertiary audiences.

8.5 Apply visual devices to help your reader skim, follow, and refer back to your discussion.

Specific fonts etc

If electronic, consider hyperlinks

Discussion:

Coherence devices work *only* when your document passes both the content and organization tests. The devices reinforce your logic for the reader. If your document suffers from irrelevant content or poor organization, coherence devices only accentuate those flaws.

Edit for coherence first. If readers can't follow your thoughts, they can't go further—even if you excel in your other editing techniques. Also, editing for coherence helps you discover any gaps in logic as you emphasize your logical flow. A document with zero coherence devices would look like one large, reader-unfriendly block paragraph. Coherence devices make documents easy to use.

Remember that as your subject matter increases in technical difficulty, your reader needs more help from coherence devices.

Technique 8.1	**Use key words in titles, subheads, and throughout your document.**

Tips:	Keep titles short, using key words. Key words name the precise topics your document discusses.
	Use key words in the subject line of letters and memos, in the purpose statement, and in the first sentence of paragraphs.
	Add value with headings. Generic headings like *Introduction, Purpose,* and *Conclusion* don't tell the reader much.
	Tie graphics to your text with captions that use key words.
	Use headers and footers to help readers keep track of their place in the document.
Warning:	Do not shift key words. If you write about a *procedure,* don't shift to *process,* then *method, approach, scheme* just for variety. If you shift words, the reader expects you had a **real,** not a *stylistic* motive.
Error Trap:	Shifting words often indicate fuzzy thought. Re-apply the Content Test.
Example:	*Score Variance* We computed the statistical *variance* between the students' *scores* on each *test.* This *variance* measures how much the *scores* cluster for a given *test.* The higher the *variance,* the greater the spread of *scores.*
See also:	Purpose statement; Revision—content test; Clarity—specific and concrete words.

Discussion:

In school we learned to use a thesaurus to avoid repeating words. However, in business and technical writing, repeating key words is a virtue. So give your thesaurus a rest. Impress your reader with your specific and concrete detail, not your vocabulary.

When you shift key words, your writing becomes inconsistent, and consequently less coherent. Keeping words consistent is more difficult when writing in a collaborative effort, especially for abstract topics such as computer applications. Whenever subject matter is abstract, technical writers must help the reader by keeping key words consistent.

Exercise: Shifting words make this memo written by a software engineer difficult to follow. Improve coherence by identifying shifting words and replacing them with consistent key words. Hint: We italicize shifting words. For one set of shifting words, we recommend a consistent key word. (Answer A-16)

Subject: ReportMaker Database *Concerns*

These *thoughts* are for your information and comments. I tried to define some *issues* we've been grappling with as we focus on the *end user*'s needs. I met with Kathy Barlowe on May 10 to discuss some of the *questions* involved in supporting *commercial graphics packages* with the database. Kathy and I tried to better define *operators* who may call for *capability* to incorporate graphics into Report-Maker.

Software Compatibility *Issues*
If we plan to offer this *ability* to the *user*, we need to test this *feature* for *standard software packages* and list guidelines for the *customer*. For example, if only a few *graphics packages* can be pulled into ReportMaker, we should say so to take out the guesswork.

concerns, thoughts, issues, questions — Make the key word *concerns*.

Exercise: Improve this excerpt from a technical report by identifying shifting words and replacing them with consistent key words. (Answer A-17)

The Office of Public Works needs to implement a fax solution that enables employees to send and receive faxes from and to their computer workstations. Public Works also wants workers to be able to share faxes among peers electronically.

Fax Solutions
We can use one of the two broad methods of implementing PC-based fax capabilities in mid-sized offices given our moderate fax traffic. The table below illustrates the two approaches and their pro's and con's:

Method	**Advantages**	**Disadvantages**
Dedicated Modem	Handles large fax volume	Very expensive
	Provides flexibility	No central control
Fax Servers	Provides central control	Requires dedicated PC server
	Shares resources	Increases network traffic

Because of the expense associated with providing a fax modem per user, we do not recommend that course of action, but prefer the server solution.

Technique 8.2   **Group your ideas in paragraphs and vertical lists.**

Tip 1:   Put the key sentence at the beginning of the paragraph, and make all other sentences logically develop the key sentence.

Use transition words (such as *therefore, consequently, also, on the other hand*) to relate thoughts.

Use the last sentence in your paragraph to summarize, draw conclusions, or lead to the next paragraph's thought.

Warning: Do not bury the key sentence in the middle of the paragraph, or omit it altogether, forcing the reader to infer the correct meaning.

Do not place the key sentence at the end of the paragraph unless you want to lead your reader through your thought process before getting to the point—sometimes used as a persuasive strategy for a hostile audience.

Do not move sentences around randomly in paragraphs. In logically developed paragraphs, each sentence leads purposefully to the next.

Example: We designed the instructions in this manual to reach the most inexperienced users. Therefore, set aside any fears about personal computers while you read this manual. Then, you may surprise yourself by becoming a computer enthusiast. Good luck and happy computing!

See also: Sentence outline; Revision—content test.

Discussion:

Your sentence outline provides key sentences to develop in paragraphs. A reader can skim your discussion by reading the first line of each paragraph.

Use transition words to explain where your thoughts go within and between paragraphs:

- forward or backward in time or discussion—*next, previously, as mentioned. . .*
- supporting or summarizing —*for example, in addition, overall. . .*
- contradictory—*however, although, in contrast, despite. . .*

Avoid transition words with multiple meanings. To avoid confusion, use *while* and *since* only when referring to time passing—do not use *while* when you mean *although*, or *since* when you mean *because*.

illustration

Exercise: The key sentence tells the reader the purpose of the paragraph. A coherent paragraph puts the key sentence first for emphasis, then logically develops it with following sentences. Why are these paragraphs incoherent? (Answer A-18)

Paragraph 1:

The TAKKS-C architecture is similar to TAKKIMS in that the local area networks are comprised of fiber rings, although the TAKKS-C fiber optic receivers are single fiber devices and less fault-tolerant than the dual port transceivers proposed for TAKKIMS. Another distinct difference between TAKKS-C and TAKKIMS is the interconnection of the sites. TAKKS-C uses ports on the host computer to interconnect geographically separated LANs. In TAKKIMS, gateways attached directly to the fiber optic LANs will provide a network that appears to the users and software as a single logical network. Unlike TAKKS-C, TAKKIMS will provide fiber optic LANs at all remote sites.

Paragraph 2:

CATS Representatives respond to telephone close requests by either closing the Customer's account or referring the Customer to the Customer Retention department. The Flag field on MBNAIS indicates to the Representative if the Customer is profitable and worth saving. If the account is profitable and the call is taken during Customer Retention's operating hours, the CATS Representative will transfer the Customer to Retention. If the Customer Retention is closed, the CATS Representative will attempt to save the account by explaining the benefits of the MBNA card and possibly offering a different interest rate (associate bank charge.) If the account is not profitable, the CATS Representative will close the account and make appropriate monetary adjustments.

Tip 2:	Use vertical lists to group information: logically related items, statements, commands, or questions.
	Be sure list entries have similar importance.
Warning:	Do not write long, haphazard lists. Explain the logic of the group as you introduce each list.
	Do not write a list when you need to write a paragraph. Paragraphs develop ideas, while lists merely group them. If you need to explain the connection between one idea and the next, write a paragraph, not a list.
Error Trap:	If a list item exceeds four lines, you may be trying to develop an idea in a list—use a paragraph instead.

Example:

Shop safety requires that all plant workers recognize the three klaxon warnings and know how to react. The klaxon warnings and corresponding reactions are described below:

1. *Continuous horn blast* means evacuate plant immediately.
2. *Three long horn blasts* mean prepare to stop line; step behind red line.
3. *Continuing one-second horn blasts* mean line-speed changing.

See also:	Coherence—transition words; Clarity—parallelism; Economy; Punctuation.

Discussion:

Lists call your reader's attention to key information, and signal that the information is logically related.

Number your vertical lists if the items show rank, order, or sequence. A list of bulletized items looks like a laundry list. Plus, readers have a hard time referencing bullets. For example, on this page *Klaxon warning number 3* is easier to reference than *Klaxon warning third bullet from the top.*

Exercise: Use vertical lists to logically group information for the reader. (Answer A-19)

1. Follow these steps to import line graphs when word processing. First, open an empty frame where you want to put your line graph. Then from the menu bar, choose Graphics, Figure, Retrieve. The retrieve menu appears and lists the line graphs you may import. Double-click on the desired filename. Your selected line graph appears in the frame. Adjust the position or size of the frame by clicking and dragging a frame handle. Then anchor your frame to the page by double-clicking the anchor icon.

2. Disbursement requests must be properly documented for payment processing. Each disbursement request must contain payment authorization and invoices signed by the approving department. It must also contain a receiving copy of the purchase order and other receipts such as shipping documents, plus a remittance for any merchandise returned. The disbursement system allows some unique purchases on a case-by-case basis. For example, hardware maintenance outside our standard maintenance contracts requires a copy of the estimate and final bill. Advertising requires proof-of-service such as a copy of the advertisement, the order, and authorization.

Lists

Logically related
Important
brevity
Easy to reference — reader + writer

How many? Generally, not > 7-8

Need an intro sentence that explains how they are logically related
Is there a rationale to the order?

Tip 3: Use short paragraphs to emphasize important ideas, and use long paragraphs to elaborate on ideas.

Warning: Avoid one-sentence paragraphs or paragraphs longer than fifteen lines.

Example:

Poor: Thank you for sending me materials about business writing. I already shared it with NACA Studios employees who indicate an interest in attending one of your workshops. I will call you this week to set the agenda for our meeting scheduled next Friday, March 4, 1994. However, based on the information you sent us, we may be able to make a quick decision to schedule training and thereby forgo the meeting. I enclosed the results of our internal needs assessment. You'll be pleased to see that writing training demand ranks third for employees and second for managers. Please treat our needs assessment data as proprietary.

Good: Thank you for sending me materials about business writing. I already shared it with NACA Studios employees who indicate an interest in attending one of your workshops.

I will call you this week to set the agenda for our meeting scheduled next Friday, March 4, 1994. However, based on the information you sent us, we may be able to make a quick decision to schedule training and thereby forgo the meeting.

I enclosed the results of our internal needs assessment. You'll be pleased to see that writing training demand ranks third for employees and second for managers. Please treat our needs assessment data as proprietary.

See also: Audience.

Discussion:

A combination of short and long paragraphs keeps the reader's interest. Too many short paragraphs seem choppy. Too many long paragraphs seem dense. Readers' eyes naturally go to short paragraphs, so put your most important ideas in the short paragraphs. But beware— if you have too many short paragraphs, none of them attract attention.

Make paragraphs horizontal rather than vertical. On an $8\frac{1}{2}$ x 11 page, keep your paragraphs less than 15 lines long. Use shifts in thought as a chance to break long paragraphs. Readers can process only so much information at once.

Business letters often use a short initial paragraph to get to the point and grab attention. They shift to longer paragraphs to explain and expand details. Then they close with a short paragraph calling for action.

A one-sentence paragraph states an idea but does not develop it. You may occasionally use a one-sentence paragraph for emphasis, but the idea must stand by itself.

Exercise: Break this long paragraph into shorter paragraphs. (Answer A-19)

ARNEWS is the bi-monthly newsletter of the Chief, Army Reserve. The Public Affairs office publishes the contents, current events in the Army Reserve. In addition, Public Affairs maintains the mailing list for ARNEWS's subscribers. To create an edition of ARNEWS, the Public Affairs office collects the articles that comprise the newsletter. Articles come from sources within the Army Reserve as well as outside sources contributing for publication. Public Affairs edits the articles and desk-top publishes a camera-ready copy of the newsletter. To create a mailing list for ARNEWS distribution, Public Affairs downloads from its membership database an ASCII file containing name, address, zipcode, and member ID. The ASCII file has one record for each line, followed by a carriage return. The record fields are delimited by commas. Public Affairs sends the camera-ready copy of the newsletter and a 16 BPI tape of the mailing list to the publisher. In turn, the publisher prints the newsletter, creates and affixes cheshire labels, then mails the newsletter.

Technique 8.3   **Preview your document or major sections within your document with an introduction.**

Tips:	Use the introduction to get the reader into your document. Include some or all of these seven parts of a good introduction:

At a minimum

1. feature your purpose statement at the beginning of your introduction
2. tell the reader how you organized your document, or section

If helpful to your reader, add

3. background and significance of topic
4. detailed description of target audience
5. information sources and research methods
6. definitions of key terms *or convention - ie, if Reviewer*
7. limitations of the document *Comments are in a diff fnt, define here*

Warning:	Do not introduce the subject instead of the document. Do not confuse an introduction with a summary or digest— a condensed version of your document.
Example:	See the introduction to this book (page 1).
See also:	Purpose statement; Sentence outline.

Discussion:

All documents need some introduction.

Recall your purpose statement. You used it to focus yourself; now use it to focus your reader. By placing the purpose statement at the beginning of your document, the readers knows *immediately* if they should read the document because they learn

1. what kind of document they have
2. if they are in the target audience
3. what the topic is
4. what they can do with the information

Your introduction proves just as valuable if the reader decides—based on your purpose statement—not to read further. In addition, the purpose statement sets your document's tone.

When telling your readers how you organized your document, tell them about the natural patterns of thought and any standard business format you used.

Describe limitations to manage readers' expectations. For example, a *preliminary* study or *partial* investigation prepares the reader for a less-than-definitive document.

Exercise: This introduction lacks one of the seven parts. Identify the six parts it has and the one missing. (Answer A-20)

Introduction to the Executive Support Manual

This manual has been designed specifically for Executive Support Personnel in the Product Support area of BAP, Inc. This manual assumes that the reader has a working knowledge of the Slide Presentation system in the VAX environment, or at least has access to appropriate manuals or knowledgeable persons on these subjects. Furthermore, it is most helpful if the Executive Support person has experience with Oracle software and at least a cursory knowledge of Generic Graphics. If not, consult the manuals available in the Product Services library.

Each month company executives view a set of charts, or slides, which have more current information than those of the previous month. Very basically, the new data comes in and is checked, then slides are generated, based on the updated data, and are checked. Next, the slides are indexed and organized into specific groups. Then the entire package is released for executive viewing. There are many other small steps involved, and the sheer volume of slides keeps the Executive Support person quite busy.

The next page displays the optimum flowchart layout of the monthly cycle. Each step has a corresponding chapter. Follow the cycle each month and refer to the appropriate chapter as you work through this manual.

In the text, anything you type is framed by double quotes ("..."). Do not type the quotes, merely type everything in between them. Syntax is very important. The set of characters <RETURN> stands for Carriage Return and is simply a tap on the key labeled "Return" on your keyboard. FDBA stands for Functional Database Administrator.

1. Purpose statement:

2. Organization of document:

3. Background and significance of topic:

4. Description of target audience:

5. Information sources and research methods:

6. Definitions of key terms:

7. Limitations of the document:

Technique 8.4   **Use front and back matter to help secondary and tertiary audiences.**

Tips:	Overview the document in an abstract or summary in less than two pages.
Warning:	Do not confuse abstract or summary with introduction. We've seen documents where the introduction and summary were word-for-word copies of each other—*ouch!* Do not write a digest (short version of your document) and call it an abstract or summary.

Example:

Short abstract: This paper answers criticisms against the 18-ton payload single-stage-to-orbit (SSTO) launch vehicle in contrast to the 163-ton payload National Launch Vehicle supported by NASA planners. Likewise this paper discusses space policy priorities and examines the immediate and long-term economic payback of both launch vehicle designs.

Short executive summary: The planning committee recommends building the warehouse as part of the factory. Further, they suggest building a pair of loading docks: one facing north, the other south. Demolishing the old warehouse and adding the new one to our factory plans extend the completion date 100 days and adds $2.32 million in costs. The committee calculates payback from smoother operations in less than 3 years.

See also:	Coherence—introductions; Sentence outline.

Discussion:

Front and back matter include transmittal letter, cover, title page, preface, abstract, summary, table of contents, list of illustrations, glossary or list of symbols, appendices, index, bibliography, notes.

Recall in Analyzing Audience that you write to your primary audience, but you may have a secondary or tertiary audience. Use front and back matter to help your secondary and tertiary audiences use your document. For example, if your primary audience is technically expert, include jargon and theory, but add an executive summary and glossary for less technical readers. If your primary audience has only a general knowledge of the subject, keep technical details in an appendix. Use titles, subheads, and other devices to help your secondary audience skim past sections that lack relevance to them.

Not all documents need an abstract or executive summary, but may have both in addition to the introduction. Write each as if the others did not exist.

An *abstract* serves experts who want an overview of key technical points: purpose, methods, and findings. Experts often use abstracts when researching. Many abstracts get put into databases that allow a key word search. Typically, abstracts use jargon and assume knowledge of advanced concepts.

An *executive summary* serves managers by highlighting conclusions, recommendations, and key points while providing minimal background and ignoring most technical details.

Your sentence outline provides key ideas to include in an abstract or summary.

Exercise: Which of the following two passages is an abstract, and which is an executive summary? How can you tell? (Answer A-20)

Mars and Luna Direct*

The concept of a coherent Space Exploration Initiative (SEI) architecture is defined and is shown to be largely unsatisfied by the conventional Earth-orbit assembly/Mars orbital rendezvous mission plan that has dominated most recent analysis. Coherency's primary requirements of simplicity, robustness, and cost effectiveness are then used to derive a secondary set of mission features that converge on an alternative mission architecture known as "Mars Direct."

In the Mars Direct plan two launches of a heavy lift booster optimized for Earth escape are required to support each four-person mission. This paper discusses both the Martian and Lunar forms of implementation of the Mars Direct architecture. Candidate vehicle designs are presented and the means of performing the required in-situ propellant production is explained. It is concluded that the Mars Direct architecture offers an attractive means of rapidly realizing a coherent SEI, thereby opening the door of the solar system to humanity.

Losing the High Frontier*

High launch prices, faltering investments in research and development, failure to support practical aspects of SDI, and a series of space disasters in the mid-1980s, are bumping the United States from our position as a strong second *(the Soviet Union has to be considered as being in first place)* in space. Our position could soon be taken by either Japan or France, and we may, before much longer, finish in fourth position after a European consortium. The nation that launched Viking, Mariner, Apollo, and Galileo space probes will be pushed, and pushed hard, by constantly shifting alliances among other nations with more mundane commercial interests in space, using discoveries and technologies paid for by the U.S. To meet this competition, a practical consortium to devise a blueprint for space is proposed.

*Reprinted with permission from *The Journal of Practical Applications in Space;* 2800 Shirlington Road, Suite 405A, Arlington VA 22206.

Technique 8.5   **Apply visual devices to help your reader skim, follow, and refer back to your discussion.**

Tips: Help your reader skim by visually highlighting key words, headings, and sentences.

Help your reader follow by using different fonts, white space, indentations, ruling lines, boxes, columns, and vertical lists to group and order items.

Help your reader refer back to your document with numbers, letters, footers, and headers.

Warning: Do not go overboard with font generators. Your text will look like a ransom note cut-and-pasted from the newspaper.

Example: Visual coherence devices include but are not limited to

Typography
     italics, **bold,** <u>underline,</u>
     different TYPEFACES
     list bullets, numbers, letters
     ruling lines and boxes
     icons

Layout
     indent, outdent, center, justification
     white space
     rows and columns
     footers and headers
     tabs

See also: Correctness—mechanics.

Discussion:

Visual devices include page layout, typography, and graphics. Remember that form follows function—don't add visual devices just because your desktop-publishing software allows it.

Consult a style guide or devise your own. A style guide determines visual (and verbal) standards. Professions, companies, projects, and individuals may create their own style guides. Every writer needs access to an appropriate style guide.

Limit your type styles to one serif (like Times) and one sans serif (like Helvetica). With each style, you can make Normal, **Bold,** *Italics*, ***Bold Italics***, and <u>Underline</u>. Limit yourself to five point sizes. You have fifty fonts to work with. Assign fonts to give specific clues. For example,

- Titles = 24pt Bold Times
- Subheads = 12pt Bold Times
- Body Text = 10pt Normal Times
- Captions = 10pt Normal Helvetica
- Citations and references = 10pt Italic Times

Then use your fonts only for their assigned purposes.

Exercise: Remember, a document with zero coherence devices is just one huge block paragraph. The following passage is well organized but needs visual coherence devices to make it reader-friendly. Add visual devices. (Answer A-21)

TO: Senior Management ¶ FROM: John Phelps ¶ DATE: May 3, 1991 ¶ SUBJECT: Long Distance Telephone Credit Cards ¶ Effective immediately, BAP, Inc. will switch from AT&T to TINKERBELL for all travel credit card calls. Employees traveling on BAP business are eligible to receive a company TINKERBELL credit card. All AT&T credit cards should be returned to Finance & Accounting (3rd floor) by C.O.B. Friday, May 7, 1991. We will close the AT&T account at that time. ¶ The TINKERBELL system requires a different procedure for using the credit card. • First, dial the TINKERBELL Travel Number, 1-800-555-4311: you will get a voice prompt asking for your code. • Enter your four-digit authorization code, then your two-digit travel code, which are printed on your card. • Second, you will enter area code and the telephone number that you want to reach. • These instructions are also printed on the front of your credit card. ¶ Each business group has its own four-digit authorization code and travel code. Calls will be charged against your group's overhead. Finance and Accounting will provide group managers with itemized call sheets so they can track their long distance credit card expenses. Separate cards can be issued for specific contracts if necessary. ¶ I am your principal contact for the new credit card system. I will assign authorization codes and travel codes. If you have problems dialing a number and have questions about procedures, contact me. If your credit card is lost or stolen, contact me or TINKERBELL, Inc. immediately.

Step 8. Edit for Coherence

Final exercise: Apply verbal and visual coherence devices to help readers skim, follow, and refer back to the following paper. (Answer A-22)

Merged Account Legislation: Public law 101-510, enacted November 1990, changed accounting and reporting procedures for Merged accounts. The legislation eliminated merged accounts, and replaced them with a revised definition of expired accounts and the newly created closed accounts. This paper describes fixed accounts prior to the enaction of Public Law 101-510, how the law changed fixed accounts, and the effect of the law on BAP. Prior to Public Law 101-510, fixed accounts, which we in BAP call annual and multi-year appropriations, had three states: active, expired, or merged. Active accounts still had obligational authority. New obligations might be established during this period. Expired accounts started and lasted for two years past the expiration of an active account's obligational authority. Obligation adjustments might be made during this period, but no new obligation might be recorded. When an an active account expired, the unobligated balances returned to the Treasury surplus fund. Merged accounts began with the third year after the point of time when the obligational authority of the account expired. At this time, the obligations were grouped with all previous budget fiscal years, and the unobligated balances were grouped at Treasury to form the merged surplus fund authority. After Public Law 101-510, fixed accounts have three states: active, expired, and closed. Each state has a different meaning. Active accounts stay the same. Active accounts still have obligational authority. Expired accounts now have a five-year period following the expiration of obligational authority. (Previously, expired accounts lasted only two years.) Expired account unobligated balances do not return to the Treasury as surplus funds, but remain with the obligated balances. During this period, all funds remain for recording, adjusting, and liquidating any obligations properly chargeable to the account prior to the time balances expired. Therefore, prior year adjustments can be made. Closed accounts now replace merged accounts. Closed accounts, known as canceled accounts, begin with the sixth year after the obligational authority expires. Then, obligated and unobligated balances return to the Treasury as surplus funds.

Step 9. Edit for Clarity

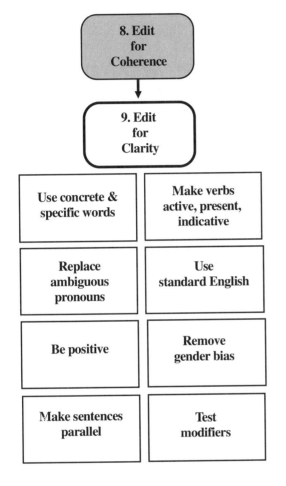

After ensuring your document's coherence, use eight techniques for clear, precise words and sentences:

9.1 Use concrete and specific words.

9.2 Make verbs active voice, present tense, and indicative or imperative mood.

9.3 Identify and replace ambiguous pronouns.

9.4 Use standard English words.

9.5 Be positive.

9.6 Remove gender bias using nine guidelines.

9.7 Make sentences parallel.

9.8 Test modifiers to make sure they're next to the thing or action they modify.

Discussion:

Make clarity your chief editorial concern. You *and your reader* want one and only one interpretation of your document. Therefore, choose words carefully, and construct sentences carefully.

Never sacrifice clarity for other editorial virtues such as economy. Sometimes you must use more words to express your point clearly. For example,

Remove the old finish.—*Short, but vague.*

Scrape or sand off the four layers of old paint.—*Longer, but clear.*

Technique 9.1	Use concrete and specific words.

Tips: Circle abstract and general words. Change to concrete and specific words.

Warning: Don't let abstract and general words communicate vague thoughts. Abstract and general words leave the impression that the author doesn't know the subject well.

Example:

| Abstract-general | The artist applied unusual colors to the surface. |
| Concrete-specific | Andy Warhol splashed hot pink and olive drab on the brick wall. |

| Abstract-general | Any help in this matter would be appreciated. |
| Concrete-specific | Please deduct the $20.16 from our account before your next billing cycle. |

| Abstract-general | The crowd approved their team's success. |
| Concrete-specific | The Washington fans roared at their Redskins' 17-to-14 victory over the Eagles. |

| Abstract-general | Recently, we have encountered problems in the mailroom. |
| Concrete-specific | In August, the mailroom misrouted 200 pieces of first class mail. |

See also: Readability.

Discussion:

Abstract words (such as *purpose, basis, matter*) are intangible; consequently, they don't communicate the same meaning to different people.

General words (such as *late, expensive, various, several*) are imprecise; consequently, they leave the reader guessing.

Concrete words appeal to your senses: sight, hearing, smell, touch, and taste.

Specific words include names, dates, places, and measurements.

Your authority as writer comes from your concrete and specific details, not your vocabulary, education, or job title. Readers assume if your content lacks concrete and specific details, you don't know the subject.

Concrete and specific details make business and technical writing more clear, lively, and interesting.

Exercise: Circle the abstract and general words. Suggest concrete and specific words. Use your imagination. (Answer A-23)

Our office has been in communication with your office recently regarding a question about issuing a refund for the remaining portion of your lease on some equipment. There are several reasons why a refund of the whole remaining portion fails to satisfy various material conditions of the lease. However, your office may contact our department to negotiate some amount suitable.

Exercise: Write two concrete or specific words for each abstract or general word. (Answer A-23)

1. transportation   _____   _____

2. subsidy   _____   _____

3. fast   _____   _____

4. contact   _____   _____

5. consider   _____   _____

6. move   _____   _____

7. benefit (noun)   _____   _____

8. familiarize   _____   _____

9. various   _____   _____

10. response   _____   _____

Technique 9.2   **Make your verbs active voice, present tense, and indicative or imperative mood.**

Tip 1:	Use active voice. Make the subject of your sentence act. Use people as subjects of verbs to add vigor.
	Eliminate passive voice. (A passive subject does not act, but is acted upon.) Circle "to be" verbs and evaluate each. Does the sentence beg the question *who* or *what* performed the action? If so, simply answer the question, and state the answer in your sentence by making the *who* or *what* the subject of the sentence.
Warning:	Do not confuse passive voice with linking verbs. A "to be" verb often links a subject with a noun restating it or an adjective describing it. In other words, the "to be" verb acts like an equals sign. Joe *is* honest. Joe = honest.
	Do not confuse passive voice with progressive tense. In this example, the "to be" verb acts as a helping verb to show the time of the action. RAMCO *is hiring* twenty new store managers.

Example:	Poor:	The settlement is agreed to. (*by whom?*)
	Good:	The principals agree to the settlement.
	Poor:	This case can be viewed three ways. (*by whom?*)
	Good:	We can view this case three ways.
	Poor:	The report is automatically generated. (*by what?*)
	Good:	One-Rite automatically generates the report.

See also:	Economy—empty verbs.

Discussion:

Voice tells reader whether the subject acts or is being acted upon.

Active voice   Michelle wrote the report.— Subject *Michelle* acts.
Passive voice   The report was written.—Subject *report* was acted upon.

When the subject is being acted upon, the reader often pauses to ask the question: *acted upon by whom or by what?* Then the reader guesses—often incorrectly.

Use active voice 90% of the time to achieve clarity and vigor.

Use passive voice when you want to de-emphasize the actor, because the actor is unknown, unimportant, or because the actor would be embarrassed.

- Unknown Actor—My umbrella was stolen.
- Unimportant Actor—Coffee and donuts will be served after the meeting.
- Embarrassed Actor—Your account was improperly debited $20,000 instead of $20.

Exercise: Identify each italicized verb as active or passive. Convert passive verbs to active. Add a subject by answering the question *by what?* or *by whom?* (Answer A-24)

1. *It is pointed out* in the article that 13-column spreadsheets *help* bankers organize their information.

2. Errors that I *have been making* for years *are* now more easily *seen* when I *edit*.

3. Minor errors *can be eliminated* through careful editing.

4. The writing *was done* by a team of experts.

5. Little attention *is being paid* to that advertising.

6. The client *was invited* by us to review the proposal.

7. The verification and validation tests *will be conducted* after the terabyte of Landsat data *is loaded* into the database.

8. The insurance investigation *is started* only after a legal complaint *has been submitted*.

9. After the contract *was won*, we *met* the client to determine how the deliverables *would be accepted*.

10. Sorry—your money *cannot be refunded*.

Use the form which is found in Appendix A ⟹
Use the form in Appendix A

The drug is designed to dissolve in water ⟹
The drug dissolves in water.

71

Step 9. Edit for Clarity

Tip 2:   Use present tense whenever possible.

Warning: Do not shift tenses without cause.

Example:

Poor: The technical approach will rely on the OS/9 operating system.
Good: The technical approach relies on the OS/9 operating system.

Poor: We have received your resume.
Good: We have your resume.

Poor: I am enclosing a copy of our audit.
Good: I enclose a copy of our audit.

See also: Economy.   *[handwritten: uncertainty. Change to present]*

[handwritten: future tense - contains]

Discussion:

Tense tells the reader when the action occurs. Present tense is clearest. Save other tenses for when you must emphasize the time of the action.

Even if the action happens in the future, you can describe the action in present tense. Consider cookbooks, which discuss future steps all in present tense.

Tenses:

present	I write.
past	I wrote.
future	I will write.
present perfect	I have written.
past perfect	I had written.
future perfect	I will have written.
present progressive	I am writing.
past progressive	I was writing.
future progressive	I will be writing.
present perfect progressive	I have been writing.
past perfect progressive	I had been writing.
future perfect progressive	I will have been writing.

Present perfect indicates an action is completed *at present*.

Past perfect indicates an action was *completed before* another action. For example, *The office had closed when we arrived.* *[handwritten: ie, to relate a actions that both occurred in the past —]*

Progressive tenses indicate actions in *progress*. *[handwritten: ie, it's ongoing + we will reach a conclusion]*

72

Exercise: Put all verbs in present tense. (Answer A-25)

User Manual for Lawnmower

You will need to follow these instructions to operate your lawnmower:

1. First, you will check the oil and gas levels.

2. Then you will ensure no debris is near the blades when you will start the motor.

3. You will next put the choke to the red line as we have shown in figure 2.

4. After you have grabbed the deadman lever with one hand, you will pull the starting rope with the other.

5. If your mower is starting with difficulty, you will need to prime the carburetor.

6. After you have finished mowing your lawn, you will clean grass cuttings from the engine area.

Exercise: Put all verbs in present or past tense. (Answer A-25)

We have received a copy of your resume in which you say you will be available to begin work in September. I have taken the liberty of forwarding your resume to our Pawley's Island project officer, Edwina Smith. I am also sending her a copy of the application you filled out when you had come for your initial interview.

You can expect the following sequence of events. Edwina Smith will return in two weeks from a bidders conference held this week. Then, she will call you, and you and Edwina will decide if you two will conduct your second interview at the Pawley's Island site. I have been thinking that you will match her requirement for a hydrologist, and that you will enjoy the Gulf Stream Project. We will pay your travel expenses.

I am looking forward to hearing if you will take the hydrologist position at Pawley's Island. Call me in three weeks if you have not heard from Edwina.

should: implies it's not mandatory, just optional — Conditions exist where you don't need to obey

"might": it's possible
Can : is possible

shall: legally obliged
will: implies consent

| **Tip 3:** | Use imperative or indicative mood. |

| **Warning:** | Do not use the subjunctive mood unless you want to create "waffle" room by introducing ambiguity. You serve yourself and the reader better by stating your conditions in the indicative mood. |

| Poor: | We should finish painting unless it rains. |
| Better: | We expect to finish painting unless it rains. |

Do not use the subjunctive mood to be polite.

| Poor: | You should plan to attend—*polite but ambiguous.* |
| Better: | Please plan to attend—*polite and clear.* |

| **Example:** | |

Indicative:	We eat our broccoli.
	States a fact.
Imperative:	Eat your broccoli.
	Issues a command.
Subjunctive:	If I were President, I would eat my broccoli.
	Expresses a hypothetical statement.

| **See also:** | Clarity—parallelism. |

"You must submit more data so that we can complete our review." "Must" is ok because we haven't invoked the Act — just modifies the clause

---

Discussion:

Verbs express one of the following moods:

Imperative mood	issues a command
Indicative mood	makes a statement of fact or asks a question
Subjunctive mood	expresses a conditional, imaginary, or hypothetical statement

Whereas imperative and indicative moods are clear, the subjunctive mood, with its *would, could,* and *should,* creates ambiguity. The verb "to be" uses the form *were* in the subjunctive mood: *"If I were king. . ."* Often, the reader can't tell whether you intend your message to be conditional, imaginary, or hypothetical. Therefore, the reader must guess. For example, is the subjunctive statement *If you were two feet taller, you could play basketball in the NBA* conditional, imaginary, or hypothetical?

Don't make the reader guess. Avoid imaginary and hypothetical statements altogether. Make your conditional statements unmistakably conditional by writing in the present tense and using terms like *if, then, else, when, may,* and *might.*

| Poor: | Should you drink, you would not drive. (Ambiguous subjunctive mood.) |
| Good: | If you drink, don't drive. (Clear conditional.) |

Exercises: Change subjunctive mood to indicative. (Answer A-25)

Dear Joe Palmer:

This letter should clear up a misunderstanding about proposed changes to our purchasing policy, which would affect your department.

It would appear that you have taken our first draft guidance too literally. We should have warned you that you had an early draft. Then possibly we could have saved this misunderstanding. Whereas the first draft said vendors should not expect to be paid in less than 600 days, the final policy states vendors should not expect to be paid in less than 60 days.

You should note other changes in the final draft of our purchasing policy. We would be grateful if you would address further questions about the new vendor policy to Mr. Smith, (991) 555-1234.

Change these subjunctive sentences to make them clearly conditional.

1. You could consider buying a house if you get a raise.

2. Should an employee be the cause of a late report, you should take disciplinary action.

Make these polite but ambiguous requests clear.

1. You shouldn't feed the animals.

2. Patrons should not use flash photography in the planetarium.

Technique 9.3   **Identify and replace ambiguous pronouns.**

Tips:	Circle *this, it, that, there,* and other pronouns.
	For the words *this, these, those, that,* ask the question *what?* This *what?* These *what?* Those *what?* That *what?* Put the answer in your sentence.
	For your *There are* and *It is* openings, ask *What* are? *What* is?
	For other vague pronouns such as *it,* look left to the nearest noun. If the *it* doesn't refer to the noun to the left, change the pronoun *it* to a noun.
Warning:	Do not use *it is, there is,* and *there are* openings.
	Do not let casual speech habits (*that's that, this means that*) creep into your writing.
Error Trap:	If you don't know what noun replaces "this," go back to Revision; evaluate content. Your assertion may be faulty.

Example:

Poor:	At present, *there are* no plans to open a third store.
Good:	At present, *we have* no plans to open a third store.
Poor:	*This* means we must store old files in the attic.
Good:	This *new load of paper* means we must store old files in the attic.
Poor:	*It*'s not worth *it.*
Good:	*The limited increase in machine speed* is not worth *$5,000.*

See also:	Coherence—key terms; Economy.

Discussion:

Ambiguous pronouns—*this, these, those, that, there, it*—make poor subjects for sentences.

Writers often use pronouns to avoid repeating nouns. However, when the pronoun doesn't obviously refer to the noun to its left, confusion sets in. For example, "Because the firm neglected the project, it failed." What failed? the firm? the project? Do not sacrifice clarity to avoid repetition.

Replace ambiguous pronouns with concrete and specific nouns.

Exercise: Circle the ambiguous pronouns and suggest changes. (Answer A-26)

Pennsylvania Department of Transportation Letter:

Dear Mr. Sparge,

For a road to be accepted into the Secondary System of State highways, it must meet two criteria.

First, it must serve three or more separate households. This is frequently referred to, somewhat misleadingly, as the "three driveways" requirement. If your local road does indeed provide exclusive service to three households, it is eligible for addition to the system.

Second, the road must meet state construction standards and it must be in good condition. That means if work needs to be done for the road to meet this, it should be financed by either the residents or the original developer before it is admitted as a standard subdivision street.

If state funds are needed, it is possible that the Rural Addition process may be used. This allows substandard roads to be taken into the system and public money used to repair them. Frequently, a special tax assessment is levied on residents to assist in paying for this. It appears that your local road may qualify for consideration under the Rural Addition process.

If you want more up-to-minute information on this, please contact me at (800) 555-1234. Thank you again for bringing this to my attention.

From a specifications document:

The learning curve for the Foxglove report generator may take up to a week. If the application has simple report requirements, it may be more efficient to hard code the report. However, after one becomes familiar with Foxglove, it becomes fairly easy to include it in applications.

From the TDR Queue window, choose the desired item from the list box by clicking on it. This highlights the item.

Technique 9.4   Use standard English words.

Tips:	Replace Latin and other non-English words with simple English.
	Use standard English words currently understood by most business and technical professionals.

Warning: Do not supplement English words with non-English words that add no meaning.

Poor: We need supplies *such as* pens, paper, *etc.*
Good: We need supplies such as pens and paper.

Don't invent or use slang.

Poor: We study the *causation* of weather *shiftivity* and the *effectability* on farms, *profitwise.*
Good: We study the cause of weather shifts and the effects of weather shifts on farm profits.

Example: Ask a roomful of professionals what *i.e.* means, and you get as many as five answers—often vigorously defended. The abbreviation *i.e.* stands for the Latin *id est*, or English *that is*. Ask the same roomful of professionals what *that is* means and you get one answer.

Writers invent words by adding suffixes such as invent*ability*, acceptance*wise*. We also add prefixes to create another class of nonstandard words. Regardless becomes *ir*regardless or even *disir*regardless.

See also: Economy—redundancy.

Discussion:

Latin and French business terms are carry-overs from the 19th century when most professionals knew these languages. Today, most people don't know what the foreign expressions mean. English is tough enough with its 1.2 million words.

As technology expands, so does our working vocabulary. Use recognized technical jargon for the appropriate audience, but avoid inventing non-standard words that do not communicate clearly.

Exercise: Match non-English or non-standard phrases with one of the standard English phrases listed on the right. (Answer A-27)

1. et al	a. to this (purpose)
2. et cetera	b. for example
3. e.g.	c. by the fact itself
4. i.e.	d. appropriate
5. ergo	e. consequently
6. ad hoc	f. by the day
7. vice versa	g. and others
8. laissez faire	h. according to
9. a la mode	i. by way of (not by means of)
10. in situ	j. that is
11. via	k. on site
12. vis-a-vis	l. adverb form of concept?
13. ipso facto	m. noun meaning department?
14. per	n. position reversed
15. in lieu	o. in the fashion
16. per diem	p. instead of
17. non sequitur	q. face-to-face
18. apropos	r. and so forth
19. departmentality	s. let people do
20. conceptwise	t. it does not follow

Technique 9.5   Be positive.

Tips: Change a negative statement to a positive statement
                  to improve clarity. Remove the *not*'s, *un*'s, and *anti*'s to
                  express the thought positively.

Warning: Do not confuse a negative statement with a negative message.

                  Negative statement of a positive message—
                  His excellent efforts did not go unnoticed.

                  Positive statement of a negative message—
                  His tawdry efforts bored us.

Example:

        Poor: It is not impossible that we made a mistake.
        Good: It is possible we made a mistake.
        Better: We possibly made a mistake.

        Poor: We are not going to meet the deadline.
        Good: We expect to deliver the manual one week late.

        Poor: Jack lives not far from the airport.
        Good: Jack lives within a mile of Dulles Airport.

        Poor: The Senate was unreceptive to the President's budget.
        Good: The Senate rejected the President's budget.

See also: Economy; Correctness—adjectives and adverbs.

Discussion:

Lawyers often use negative statements to create some ambiguity. Business and technical writers often use negative statements to beat around the bush when delivering a negative message.

Positive statements persuade better than negative statements. Change "Don't take your coffee break before 10:00 A.M." to "Take your coffee break after 10:00 A.M."

Exercise: Change these negative statements to positive statements. (Answer A-28)

1. Your problem is not being unable to justify the need for a company car, but rather obtaining the necessary funding.

2. I don't find anything the slightest bit unsobering in those dealings.

3. The Court rejected an appeal by Maryland officials challenging a state law allowing the investment tax credit.

4. We can't hear you.

5. If the operator does not remove the red safety tag from the disk drive, the system will not boot up and software installation cannot continue.

6. The General Partners will not be found liable for non-performance unless the General Partners cannot show that the Limited partners failed to prove that the General Partners did not act in good faith.

7. We were unable to accept your bid because we were unable to justify your higher costs for services not unlike those offered in a less costly bid.

8. It is not an uncommon practice for employers to give employees bonuses at Christmas.

9. I do not intend to appear unreasonable.

10. Her response was not illogical, but it did not include key information.

Technique 9.6	**Remove gender bias using nine guidelines.**

Tips:	Use the following nine guidelines to remove gender bias.
	1. Use a plural noun and eliminate the use of the gender pronoun.
	2. Reword to eliminate the gender pronoun.
	3. Substitute *person* for *man* or *woman*.
	4. Substitute *one* or *you* for *he* or *she*.
	5. Write *he* or *she*, *his* or *her* (not *he/she*, *his/her* or *s/he*).
	6. Avoid unnecessary references to marital status.
	7. Be consistent when referring to people by last names and titles.
	8. Replace occupational terms ending in *-man* or *-woman* with another term.
	9. Use synonyms.

Warning:	Avoid the ridiculous constructs like *personhole cover* instead of *manhole cover*.
	Do not use plural pronouns to replace singular nouns.

Poor:	Every employee selects their own health care plan.
Good:	Every employee selects his or her health care plan.
Or:	Every employee selects a health care plan.

Example:

Poor:	This model home is a man's ambition and a woman's dream. Mom will love the gourmet kitchen while Dad enjoys the large hobby room.
Good:	This model home is your ambition and dream. Your family will love the gourmet kitchen and large hobby room.

See also:	Correctness—grammar and mechanics.

Discussion:

Gender bias occurs when language stereotypes or unnecessarily distinguishes people by gender. We can take advantage of our large vocabulary, gender neutral plurals, and gender neutral second-person pronouns to eliminate much of English's bias.

Remove gender bias from your message to avoid these problems:

Practical—Some of your *firemen* are really female firefighters.

Economic—"A family *man* needs insurance" excludes half of the market.

Legal—"We seek a professional who has *her* MBA" illegally excludes male applicants.

Exercise: Think of a suitable synonym. (Answer A-28)

1. businessman	6. stewardess
2. craftsmanship	7. fatherland
3. foreman	8. gentlemen's agreement
4. middleman	9. salesman
5. sportsmanship	10. waitress

Exercise: Edit these sentences to avoid gender bias. (Answer A-29)

1. Experienced waiters make dining more pleasant.

2. The average American drives his car every day.

3. A man who wants to get ahead works hard.

4. If a man plans ahead, he can retire at age 60.

5. Each senator selects his staff.

6. Be sure to bring your husband to the D.C. Armory Flower Show.

7. Reagan, Gorbachev, and Mrs. Thatcher dominated politics in the 1980s.

8. The fireman and policeman controlled the crowd.

9. A homeowner can deduct interest expenses from his taxes.

10. The user can make only three attempts to enter his password before the machine locks him out.

Technique 9.7	**Make sentences parallel.**

Tips:
Make sentences and parts of sentences grammatically parallel when they are in pairs, series, and vertical lists.

Pay special attention to sentences that have *either. . .or. . ., not only. . .but also. . ., between. . .and. . .*

Keep mood and tense parallel in logically equal sentences. Use your ear—most writers recognize a shift in parallelism when they hear it.

Warning:
Do not confuse the reader by making unnecessary changes in verb tenses and mood.

Error Trap:
If you have a difficult time making a list item parallel, consider the strong possibility that the list item does not logically belong to the group.

Example:

Poor: Our salesroom is clean, comfortable, and has a lot of space.
Good: Our salesroom is clean, comfortable, and spacious.

Poor: Their firm leads not only in production, but also is a leader in marketing.
Good: Their firm leads not only in production, but also in marketing.

Poor: With Weatherby software the user can get a weather report, produce reports, zoom in on an area for detail, and the screen displays current aerial photos.
Good: With Weatherby software the user can get a weather report, produce reports, zoom in on an area for detail, and view current aerial photos.

See also:
Correctness—grammar and punctuation; Coherence—vertical list.

Discussion:
Parallelism refers to using the same grammar for logically similar sentences and parts of sentences. Parallelism helps the reader group information for easier understanding and recall. For example, express a series of instructions in imperative mood: *Put up the signs, arrange the tables, and display the new merchandise.*

Exercise: Make these sentences parallel. (Answer A-29)

1. The choice between an optimum system design or one that is less than desirable is affected by our R&D budget and how we use commercially available software.

2. Management will assess your job performance by the following criteria. Are you neat and well-groomed, did you get your assignments done on time, have you been flexible and are you willing to learn.

3. Our latest magazine issue will lose money because we did not fill the advertising space, we needed 2,000 extra copies for promotion, and we pay too much for paper.

4. According to the request category, we will either recommend the zoning board approve the plan outright or the review committee request more information.

5. When you make the list, arrange the items in order of importance, write them in parallel form, and all the items should be numbered.

6. We propose the following agenda for the meeting:

>   a. Calling the meeting to order
>   b. Set date for next meeting
>   c. Taking the roll call
>   d. Election of new officers

7. The tax committee voted to

>   review the materials being purchased for the tax library
>   submit a report on new billing rates
>   client development programs
>   annual tax department party

8. By next Monday, please complete the research, analyze the various positions, and you should hand in the report.

9. The new accounting software package fails to meet our requirements for several reasons:

>   1. It is too slow.
>   2. Do the menus need to be so complicated?
>   3. Should the power fail, it would lose all my data.
>   4. Enter a future date, and the ledger will not balance.

10. When you build your database, either use dBase IV for Windows or Altbase for OS/9.

Technique 9.8	**Test modifiers to make sure they're next to the thing or action they modify.**

Tips:	Find the modifier. What does it modify now? What should it modify? If misplaced, put the modifier in the right place. If dangling, give the modifier something to modify.
Warning:	Do not let imprecise speech patterns creep into your writing. Pay special attention to the modifiers *only* and *almost*.
Poor:	I *only* have eyes for you. —song by A. Dublin and H. Warren (*only* I, because you're ugly? *only* eyes, because you stink?)
Good:	I have eyes for you *only*.

Example:

Misplaced:	*As the main ingredient for homemade chili*, Mr. Brody's wife selects the finest cuts of meats. (Mrs. Brody wants that modifier moved after *meats*.)
	I am happy to report *in April* no accidents occurred. (Was April accident free? Or was the report in April?)
Dangling:	The local butcher can recommend the best cuts of meat *to make a good chili*. (Who's making the chili? the meat? the butcher? Give the modifier something to modify: *so you can make a good chili*.)
See also:	Correctness—grammar.

Discussion:

Modifiers are words we use to describe, limit, or otherwise alter the meaning of other words in a sentence.

Misplaced modifiers and dangling modifiers confuse the reader.

- A *misplaced modifier* is a word, or group of words, in the wrong place, usually too far from the words you intended to modify.
- A *dangling modifier* cannot logically refer to any other word in the sentence. Usually, the word it was supposed to modify is missing. You can repair a dangling modifier by giving it something or someone to modify.

Exercise: Correct misplaced modifiers. (Answer A-30)

1. Our department only receives a limited amount of money to spend on office equipment.

2. The notices for the employees' benefits meetings were all published in the employee newsletter along with e-mail sent to our field offices.

3. The Fish and Game Club announced tuna are biting off the west coast.

4. I allowed the staff to take a day off before I took my vacation without pay.

5. A recent White House report was released claiming that acid rain is linked to methane emissions by the President's scientific advisor.

6. For your information, I have enclosed our company's financial statements.

7. Senior managers will meet with the Chairman and CEO at 10:00 Tuesday morning about the Competition in the board room.

8. Soaring high above the clouds, we watched the space shuttle fly into space.

9. The Department of Housing only receives limited funds for maintenance.

10. Two campers were found shot to death by the park rangers.   *Can be fixed by eliminating the passive voice*

11. I am happy to report, in April, 18 districts reported no lost-time accidents.

12. NL Inc.'s goal is not to report any accidents, but it is extremely important to reduce and eliminate lost-time accidents.

13. I met a man with a wooden leg named Smith.

14. Please mark your calendars on April 5 for the annual tax meeting.

Exercise: Correct dangling modifiers. (Answer A-30)

1. To determine the final costs, the man-hours must be totaled and multiplied by the hourly rate.

2. After lying on the bottom of the Atlantic Ocean for seventy years, the photographers brought back pictures of the Titanic.

3. Having studied the client's requirements, the technical approach must include icon-driven menus.

4. Being unable to read English, our user manual could not satisfy our Korean customer.

5. To be a successful manager, good writing skills are essential.

6. After writing a sentence outline, the first draft was easy.

7. Confident of our success, the proposal went to the client.

8. After raining all day, we moved the reception indoors.

Final exercise: Vague terms, passive voice, shifting tenses, ambiguous pronouns, Latin phrases, negative statements, gender bias, an unparallel vertical list, and misplaced modifiers make this passage unclear. Edit for clarity. (Answer A-31)

Mr. Smith
P. O. Box 2456
Arlington, VA 22145

Dear Mr. Smith:

We are unable to approve your request to register your farm edifice (i.e., a barn) as a historical building. It cannot be approved because it cannot meet minimum requirements for several reasons:

1. Although the barn was originally built in 1927, by an alleged descendent of Lord Fairfax, later improvements to accommodate cows have caused the Historical Society to place the barn's effective date much later.

2. The barn requires extensive repairs that neither the owner can afford, nor the Society.

3. Should the barn be repaired at the owner's expense, the owner would still have to pay for his own insurance.

4. The barn has already been condemned to allow a two-lane roadway to be constructed.

This means that there is no way to register the old cow barn as a historical site and use that to impede construction of the new Route 230. Frankly, if it had not been condemned for the road construction, it would have been condemned as a structural hazard. Ergo, the owner would have had to pay for razing the barn instead of the state. Some consolation may be taken in knowing that it was designed to provide faster and safer transport of the many dairy vehicles that travel Route 230.

Step 10. Edit for Economy

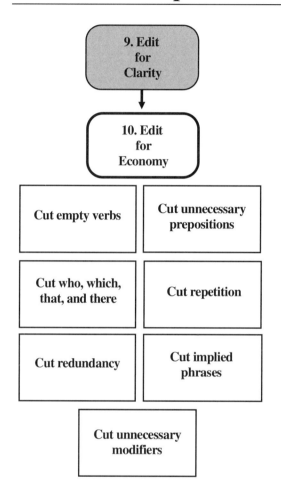

9. Edit
for
Clarity

↓

10. Edit
for
Economy

Cut empty verbs

Cut unnecessary
prepositions

Cut who, which,
that, and there

Cut repetition

Cut redundancy

Cut implied
phrases

Cut unnecessary
modifiers

After editing for clear words and sentences, you're ready to cut unneeded words. Use seven techniques to cut deadwood:

10.1 Cut empty verbs.

10.2 Cut unnecessary prepositions.

10.3 Cut who, which, that, and there.

10.4 Cut repetition.

10.5 Cut redundancy.

10.6 Cut implied phrases.

10.7 Cut unnecessary or vague modifiers.

Discussion:

Words that add no meaning to your sentence are deadwood. Deadwood obscures your message, slows down the reader, and makes your prose look lazy. Remove deadwood to make your writing clear, efficient, and vigorous. You can probably cut 20-50% of the words from your draft with no loss of meaning. Keep track of your most common deadwood habits and practice editing techniques to eliminate them.

Vague terms, ambiguous pronouns, future tense, subjunctive mood, and passive voice add to deadwood. By editing for clarity, especially by using specific and concrete words, avoiding "to be" verbs, and eliminating ambiguous pronouns, you have already cut some of the draft's deadwood.

After you methodically circle and delete deadwood, only concrete and specific words remain. If no concrete or specific words remain, you need to omit or clarify the sentence.

Technique 10.1   Cut empty verbs.

Tips: Challenge your "to be" verbs and other inactive, vague verbs. Know your own habits.

Find the more specific "buried verb" and use it as an active verb in the sentence. *— esp. nouns ending in -ment, -ation, - sion, -ance*

compliance
performance

Replace many verbs with one.
You will have to decide becomes *You must decide.*

Warning: Do not accept the first buried verb you find. Choose the most specific verb.

-able
preferable
advisable
△ to prefer, advise

Example:

Passive voice:	The system *was implemented.*
Active voice:	Tom *implemented* the system.
Linking verb:	Tom's system *is* down.
Active verb:	Tom's system *crashed.*
Buried verb:	Tom *did* the system *implementation.*
Active verb:	Tom *implemented* the system.
Passive voice with buried verb:	A *decision may be reached* through *consideration* of the facts.
Active verb:	You *may reach a decision* by *considering* the facts.
Better:	*Consider* the facts before you *decide.*

See also: Clarity—passive voice.

Discussion:

A specific verb is more clear *and economical* than a buried verb.

Buried verb is wordy.	*Specific verb is economical.*
They had a discussion.	They discussed.
They passed judgment.	They judged.
They made a recommendation.	They recommended.
They reached a conclusion.	They concluded.
They enjoyed prosperity.	They prospered.
They did maintenance.	They maintained.

Learn to recognize buried verbs by these common endings:

-ion	application—apply
-ent	statement—state
-ness	forgiveness—forgive
-fulness	wastefulness—waste
-ance	sustenance—sustain
-ble	supportable—support
-ty	totality—total

Exercise: Cut empty verbs. (Answer A-32)

1. The task was performed by the manager.

2. New theories were expounded in the supervisor's reports.

3. This was a contract written by the vice president.

4. My tonsils were removed by the doctor.

5. The team, despite its best efforts in the development stage, was forced to delay the start-up.

6. The position of the director, as reported in last week's newsletter, and commented on by many, remained precarious.

7. We made an attempt to effect a repair on the motorcycle.

8. The board of directors formed a decision to give notice to employees about this year's payraises.

9. It was the determination of the auditors that BAPCO remains in compliance of generally accepted accounting principles.

10. Design of the system was contracted to Blue Communications, Inc.

11. The following questions were raised by the clients.

12. In making this decision, you should be mindful that safety must be met at all times.

Technique 10.2   Cut unnecessary prepositions.

Tips:	Cut *of* when used to show possession. *Design of the system* becomes *system design.* Cut cliche prepositions. *In accordance with* becomes *following.* *In order to* becomes *to.* *In back of* becomes *behind.* *With regard to* becomes *regarding.*
Warning:	Do not remove the prepositions if a long noun string results. The *20-megawatt capacity of hydro-electric generators beneath the Hoover Dam.* . . doesn't improve by removing the prepositions: *the Hoover Dam's base's hydro-electric generators' 20-megawatt capacity.* Remember, don't sacrifice clarity for economy.
Example:	
Poor:	*In order* to review the files *of the* company, I traveled *to* Denver.
Good:	To review company files, I visited Denver.
Poor:	We have no vacancies *at this point in time.*
Good:	We have no vacancies now.
See also:	Economy—implied phrases.

Discussion:

Prepositions link a noun or pronoun to the rest of the sentence. English uses about seventy prepositions.

Prepositions express relations such as

direction:	from, to, into, across, toward, down, up
location:	at, in, on, under, over, beside, among, by, between, through
relative locations:	or, against, with
time:	before, after, during, until, since

Exercise: Write alternatives to these prepositional phrases. (Answer A-32)

1. Call *at about* 5 o'clock.

2. *In accordance with* company policy. . .

3. He wrote *with the purpose of*. . .

4. Submit your plan *for the purpose of*. . .

5. Put the phone *on top of* the desk.

6. She is *in the midst of* a big job.

7. *In spite of the fact that*. . .

8. He is an expert *in the area of* finance.

9. Go *in back of* the shed.

10. We are *in receipt of*. . .

11. He worked *over and above*. . .

12. *In the interest of* safety. . .

13. *With regard to* your promotion. . .

14. Indicate *as to whether or not*. . .

15. Because *of the fact that*. . .

16. Because *of this reason*. . .

Exercise: Cut unnecessary prepositions from the following sentences. (Answer A-33)

1. ~~In order to meet the objectives of this test,~~ XYZ, Inc. ~~has to~~ draw upon the expertise of ~~several people belonging to and part of~~ the staff ~~in the La Jolla Laboratory.~~

2. The review by Dr. Roger of the draft of the report by the committee is on hold until the DOD review is completed.

3. In order to qualify for the exemption from taxes to local businesses, the sales of tickets to the series of lectures must conform to each of the following requirements.

4. One of the purposes of the whole project is to ensure that the Navy receives maximum return on its data documentation efforts.

5. In making this decision, you should be mindful that safety must be met at all times.

Technique 10.3   Cut who, which, that, and there.

Tips:	Cut *who, which, that* and *there* when combined with a "to be" verb, because they add no value to the sentence.

Warning:	Do not cut *that* when used to separate two thoughts.
Poor:	Mrs. Martin explained to him her newspaper vanished.
Good:	Mrs. Martin explained to him *that* her newspaper vanished.
Poor:	Economists warn business will suffer from new tax rules.
Good:	Economists warn *that* business will suffer from new tax rules.

Example:	
Poor:	Our chief engineer, *who is* a graduate of M.I.T., personally supervised the factory design.
Good:	Our chief engineer, a graduate of M.I.T., personally supervised the factory design.
Poor:	The copier, *which is* obsolete, must be replaced.
Good:	The obsolete copier must be replaced.
Better:	Replace the obsolete copier.
Poor:	*There is* a simple solution to our cash flow problem.
Good:	A simple solution to our cash flow problem exists.
Better:	We can solve our cash flow problem simply.

See also:	Economy—empty verbs; Clarity—ambiguous pronouns.

Discussion:

Who, which, that, and *there* combined with a "to be" verb link a noun, pronoun, or modifier to the rest of the sentence. Delete the *who, which, that* and *there* plus the "to be" verb, and keep the noun, pronoun, or modifier.

For example,
Change "*Marty, who is my friend.* . ." to "*Marty, my friend.* . . ."
Change "*Marty, who is friendly.* . ." to "*Friendly Marty.* . . ."

Exercise: Cut *who, which, that*, and *there* from these sentences. (Answer A-33)

1. Ann Jones, who is the leader in our contract negotiations, wants to meet you on the six o'clock air shuttle, which you usually take.

2. Please select a desk that is more suitable to your work.

3. Work continues on the Vega Project, which is scheduled for completion next summer.

4. He added a requirement that was the same as ours.

5. The policy committee, which is composed of local elected officials from Clark County, chose not to include a request for more road salt in their final budget that was submitted.

6. Remove the red safety tag, which you'll find next to the oil drain plug.

7. I hope that this letter answers your questions.

8. There are three topics to discuss in our meeting.

9. If the customer requests statement copies that are older than six months, you must look in the microfilm library.

10. Employees must report any plant accident resulting in lost labor time to their shift supervisor, who is responsible for safety.

Technique 10.4 Cut repetition.

Tips:	Use vertical lists to reduce repetition of unimportant words or phrases.
Warning:	Do not shift words to avoid repetition. Coherence demands that you repeat key words to keep the reader on track. Give the thesaurus a rest.
Example:	Using a vertical list cuts this passage from 59 to 47 words.

Poor:

To qualify for the fifty-dollar rebate, you must accomplish four steps. First, you must fill out completely and sign the accompanying 3-by-5 card. Second, you must attach the bar-code label as proof of purchase. Third, you must enclose the original cash register receipt—no photocopies allowed. And fourth, you must enclose a self-addressed stamped envelope.

Good:

To qualify for the fifty-dollar rebate, you must

1. fill out completely and sign the accompanying 3-by-5 card
2. attach the bar-code label as proof of purchase
3. enclose the original cash register receipt—no photocopies allowed
4. enclose a self-addressed stamped envelope

See also:	Coherence—key words.

Discussion:

Repetition is not always bad. Recall from editing for coherence that you use repetition to emphasize key ideas. Avoid repeating unimportant words and phrases because you draw the reader's attention to them and away from your key ideas.

Do not shift words just for variety. Although we got extra credit for varying our word choice in school, word choice variety confuses the reader of business and technical documents. For example, if you write about a *concept*, then shift the word to *idea*, then *approach, scheme. . .* the reader thinks you are making distinctions, when in fact you are using the words as synonyms.

Exercise: Cut the repetition. (Answer A-33)

1. He added an additional requirement the same as ours.

2. Downsizing, as exemplified in the example above, is key to the minicomputer boom in business.

3. Each stock item record contains a stocking conversion factor (the stocking conversion factor being the number of end-use units contained in one stocking unit).

4. This regulation is more important than other regulations.

Exercise: Cut unnecessary repetition from this passage. (Answer A-34)

Training Conferences. We planned three training conferences for government employees. The first training conference occurs approximately 45 days after the contract award. This first training conference starts with a working meeting and review of initial planning documents and requirements documents for training. The second training conference happens at day 90 to coincide with our first deliverable, the draft AIS training and technical manuals. In the second training conference, we review customer comments of the manuals as well as the skills analysis report, plan of instruction, and course outlines. The second training conference also helps resolve any concerns before we design and develop the training courses. The third conference happens about 225 days after contract award for the review and comment of the training materials and schedules. Other training concerns will be addressed as necessary during the third training conference.

Technique 10.5 Cut redundancy.

Tips: Cut words that say the same thing different ways.

                   Cut doubling.

Warning: Do not cut words or phrases that make useful distinctions.
                   For example, We offer *efficient* and *proven* methods.

Example:

   Poor: This *regulatory rule* is more important than others.
   Good: This regulation is more important than others.

   Poor: His suit *of clothes* is *a gray color.*
   Good: His suit is gray.

   Poor: In *the month of* June, Wooster Shire condos reported the
              same vacancy rate as *in the month of* May.
   Good: In June, Wooster Shire condos reported the same vacancy
              as in May.

   Poor: The potential for error can be kept *reasonable and acceptable.*
   Good: The potential for error can be kept reasonable.

See also: Economy—repetition; Clarity—concrete and specific words.

Discussion:

Redundancy often results from using an abstract or general word that means the same as a specific or concrete word. For example, *three feet in length, the subject of chemistry. Three feet* and *chemistry* are specific. *Length* and *subject* are abstract.

Doubling occurs when writers join two synonyms or near-synonyms by *and* or *or.* For example, "We offer *effective and successful* methods." Even when the doubled words are not exact synonyms, the subtle nuances in meaning tend to confuse rather than clarify the message. For example: *Attend this urgent and important meeting* is less direct than *Attend this urgent meeting.*

Doubling is common in English. In 11th century England, when half the nobility spoke Anglo-Saxon, the other half French, they wrote key terms in contracts in both languages. Hence, *null* and *void.*

Some writers have the bad habit of doubling. Adding the second word is like dropping the other shoe. *This and that's* run down their pages. Beware of word pairs joined by *and.*

Exercise: Cut redundant words and phrases. (Answer A-34)

1. The subcontractor's concerns appear to be valid and important.

2. The determination of whether or not standard, that is regulated, procedures are required to manage a series of multiple performance tests is always subject to question and is not susceptible to a conclusive determination.

3. BAMCorp's singularly unique personnel and technical package will completely and professionally fulfill, as well as satisfy, all your complex and challenging requirements.

4. Doubling can detract from and confuse the message or idea.

5. The company ~~especially~~ wishes to recognize ~~and compliment~~ Mr. Smith for five years of ~~unselfish and~~ generous ~~aid and~~ support to the Little League Baseball program in Falls Church.

6. The clients asked these questions in their request for information.

7. From the airport #1: We want to be the first to wish you a happy and pleasant day in the Washington, D.C. area or, if you are continuing your travels, we want to wish you a pleasant trip to your final destination, wherever that might be.

8. Also from the airport #2: Please make sure your tray-tables are fastened and secure in an upright position for landing.

9. From the airport #3: Please make sure your carry-on luggage is of the type and size that can be stored in the overhead bin compartment or under the space beneath the seat in front of you.

10. Customers can access their account information by modem, that is, go on-line.

Technique 10.6   Cut implied phrases.

Tips: Cut phrases already understood from context.

                 Cut overly formal courtesies.

Warning: Do not cut meaningful qualifiers.
                 For example, *In most cases, dog bites are not fatal.*

Example:

   Poor: *The purpose of* this memo *is to* review the employee health plan.
   Good: This memo reviews the employee health plan.

   Poor: *The question arises*: "Who is most qualified?"
   Good: "Who is most qualified?"

   Poor: *It seems to me that* the global warming theory lacks proof.
   Good: The global warming theory lacks proof.

   Poor: *On behalf of the entire staff*, I *would like to* thank you for
                 your *years of* outstanding service.
   Good: The staff and I thank you for your outstanding service.

See also: Economy—prepositions.

Discussion:

Use qualifying phrases if you need to be legalistic. *"In the opinion of this court. . ."* or if you need to waffle *"In this instance, we suggest that. . ."*

However, many qualifying phrases are cliches:
   In accordance with these regulations. . .
   One may conclude that. . .
   Within the realm of possibility. . .

Implied phrases, like other deadwood, obscure your message and weaken your writing. Cut them for a clearer and stronger writing voice.

Poor: *Perhaps it would be wise to* read the instructions.
Good: Read the instructions.

Exercise: Cut implied phrases. (Answer A-35)

1. ~~As you may already know,~~ Lockheed and Martin Marietta merged to become the world's largest defense company.

2. It should be noted that these new theories mark a radical change in the way scientists view the universe.

3. All things considered, our new office manager shows promise.

4. It is suggested that you send an invoice within 30 days of completing work.

5. Please feel free to call me if you have any questions.

6. (When you find time,) please give me your decision about whether or not you want me to work late.

7. Most experts claim that children need to eat a well-balanced breakfast before going to school.

8. Before we begin our discussion, remember that these remarks are strictly off the record.

9. We at BAP Industries would like to take this opportunity to thank all our vendors for their support in our on-going activities.

10. You may establish another category for the purposes of recording, adjusting, and liquidating other obligations properly chargeable to the AIS contract.

11. Thank you for your cooperation.

12. On the report in question, please write the claim number in the blank space available.

Technique 10.7 Cut unnecessary or vague modifiers.

Tips:
Choose nouns and verbs carefully to convey precise meaning and intensity. If the noun or verb needs help, use concrete and specific modifiers.

Cut abstract and general modifiers.

Warning:
Do not try to intensify a noun or verb with vague modifiers like *very, great, extreme. . . .*

Example:

Poor: We must solve this *very* difficult problem.
Good: We must solve this difficult problem.

Poor: Ms. Jones *very carefully* read our report.
Good: Ms. Jones scrutinized our report.

Poor: Your access code does not *actually* appear on the screen.
Good: Your access code does not appear on the screen.

See also:
Economy—redundancy; Clarity—concrete and specific words.

Avoid
Several : The noun it modifies is plural — redundant
numerous
Various

Discussion:

Modifiers are words we use to describe, limit, or otherwise alter the meaning of other words in a sentence. If you changed abstract and general words into concrete and specific words, you will have already cut most unnecessary modifiers.

Many unnecessary modifiers are cliches.

- *Exact same*—different from just plain same?
- *Mutually agreeable*—what other kind of agreeable is there?
- *Most unique*—like unique, uniquer, uniquest?
- *Join together*—to differentiate from joining apart?
- *Absolutely sure*—not quite as sure as positively sure?
- *Both cooperate*—or you cooperate while I resist?
  almost positive

Exercise: Cut unnecessary and vague modifiers. (Answer A-36)

1. I usually write my first drafts very quickly.

2. On the airplane: "Please use extreme caution when removing carry-on luggage from the overhead bins."

3. The clients asked the following specific questions.

4. Actual design of the system was contracted to Blue Communications, Inc.

5. BAP Industries offers a most unique solution to your complete personal computer needs.

6. Are you absolutely sure you unplugged the coffee pot?

7. Management remains fairly optimistic that we can meet our relatively high sales quotas.

8. We greatly appreciate your outstanding support.

9. Martin's analysis was completely accurate, but his conclusion was totally wrong.

10. Sally first debuted her new innovation to the public last month.

11. Write up the meeting notes, then pass them out to the committee members.

12. The inspectors were rather reluctant to sit down during the testing while I stood up.

13. If you two will both cooperate with each other, we can all achieve our goals.

14. Pam and John find it mutually agreeable to join together in the bonds of marriage.

15. The auditors determined that BAPCO complies with generally accepted accounting principles.

Final exercise: Try to cut 25-50% of the 302 words in the following passage. (Answer A-37)

MEMORANDUM

TO: Staff

FROM: D. Rose

SUBJECT: Budget Planning Sessions

You should already be in receipt of the pages attached in the enclosure to this memo. They reflect our revision to the first and only schedule of budget planning sessions, which was issued on this past January 5. There are schedule changes regarding the session dates and times made by us to accommodate a need to include more current and timely data from our operations in the field. It was felt that the recent reorganization of offices in the field would change and otherwise affect the projections for both cost and revenues in the budget.

It is possible that additional (that is, follow-on) budgetary sessions may be added as necessary. Any such follow-up sessions will be made available for the presentation and review of projections of cost and revenue that may have required research or investigation from our offices in the field.

Participants in this process of the budget should gather, acquire, and prepare budget numbers in advance of the first session. The attached budget format should be considered and used as a guideline. The intent of the budgetary sessions is to spend as much time as required to properly cover a specific topic in as much detail as necessary.

If you have a specific interest in, or desire to attend a particular budget session, and you are not a member of a field office budget team that is meeting regularly at this time, please let me or any Finance Department team member be aware of your interest so you may be notified of any schedule variance that may affect the sessions of interest to you. With the exception of Wednesday, each daily session is scheduled to be an "all day" session, meeting from 9:00 in the morning to 12:00 noon and from 1:30 in the afternoon to 4:30 in the evening.

Step 11. Check for Readability

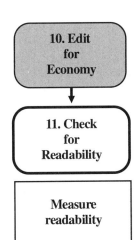

10. Edit
for
Economy

11. Check
for
Readability

Measure
readability

Use short
words

Break long
sentences

When you finish editing for coherence, clarity, and economy, use three techniques to make sure your language is easy to read:

11.1   Measure readability by determining average sentence length and percent of long words.

11.2   Replace long words with short words.

11.3   Break long sentences.

Discussion:

Readers want to read quickly and accurately. Readers resent difficult language.

Readability refers to how easy or difficult the reader finds your language. Short sentences and short words are easier to read and understand.

Writers who follow the writing system usually achieve good readability after completing the editing steps. Nevertheless, check readability after editing, and use suggested techniques to improve if needed.

Technique 11.1   **Measure readability by determining average sentence length and percent of long words.**

Tips: Use a "canned" readability algorithm or calculate average sentence length and percentage of long words—3 or more syllables. (Disregard *es* and *ed* suffixes, and capitalized words.)

For comfortable readability, keep your average sentence length less than 20 words, and use 10% or fewer long words.

Use the measurement to target your efforts to make your language more readable.

Warning: Do not use readability algorithms until after you edit for coherence, clarity, and economy.

Example: Average Sentence Length = 20   Long Words =18%

Founded in 1982, International Zymurgy has become one of the largest *suppliers* for the *rapidly expanding* micro-*brewery* market in the United States. The past 14 years' success resulted in a steady growth from a 5-person *company* working from a garage to a 825-person *corporation operating* four *regional* warehouses in Los Angeles, Chicago, New York, and Richmond. We *recently* opened a *subsidiary* office in Alberta, Canada, and an office in Mexico City, Mexico. Since we first opened our doors, we have expanded our support to micro-*breweries*. Today, we offer complete turn-key systems, over 2,000 brewing *ingredients*, plus *management, marketing*, and *operations consulting*.

See also: Clarity.

Discussion:

You can measure the complexity of language. Many wordprocessing and grammar-checking software products include readability algorithms: Flesch Reading Ease, Flesch-Kincaid Grade Level, Gunning's FOG Index.

If you don't have a "canned" algorithm, simply calculate

   1. average number of words per sentence

   2. percentage of long words—words with 3 or more syllables
     (Disregard *es* and *ed* suffixes, and capitalized words.)

Use a sample of 100+ words. Don't include vertical lists in your sample. In lengthy documents, measure readability in several places.

Remember that subject matter complexity does not determine your language's complexity. If you have a complex subject, you have a greater obligation to make the language simple. Making the complex simple is the hallmark of brilliant communication.

Write at the level your readers *want* to read, not at the level they *can* read. Even sophisticated magazines publish articles at the high school sophomore reading level.

Exercise: Measure readability by calculating the average sentence length and percentage of long words for each passage. (Answer A-38)

Managing Proposal Commitments

In order to lay a foundation to consider principal issues in managing to proposal commitments, we must first consider and accept that companies in both their government and commercial businesses, lack adeptness in managing the very commitments, and consequently the risks, to which the managers must manage.

In order for the acquisition of adeptness in the management of proposal commitments, companies must understand the imperfections in their currently used management systems. The interrelated areas in which companies seem to appear to experience the preponderance of substandard performance in managing the commitments themselves are inconsistency among commitments and ignorance of the potential negative outcomes of commitments.

Why do companies experience difficulty in managing their proposal commitments, and what effect does that have on companies' abilities in managing to those commitments? The legal community would point to suggestions from our experience adjudicating contracts that a principal reason is that the personnel who write proposals and make the commitments are not the same personnel who eventually provide management of the projects. Companies usually view proposal writing and project performance as discrete tasks rather than part of a continuum.

How to Manage Proposal Promises

To learn how to manage proposal promises, we must accept that firms lack the skills to manage their promises. Therefore, they incur risks.

To acquire the skill to manage proposal promises, firms must understand the flaws in their present management systems. Firms experience poor results in managing promises because they make promises that conflict. Also, they don't know the likely bad outcomes of promises they make.

Why do firms have problems managing their proposal promises? Contract lawyers point to one main reason. The people who write proposals and make the promises are not the same people who later manage the projects. Firms often view proposal writing and project performance as discrete tasks rather than part of a process.

Technique 11.2 Replace long words with short words.

Tips: Challenge each long (3 or more syllable) word not capitalized. (Disregard *es* and *ed* suffixes.)

Keep those that have no good short word substitute, including jargon your readers expect. Replace others with 1 or 2 syllable words.

Limit your long words to less than 10% to keep your writing at a high school level.

Trust your natural speaking voice. Long words sound unnatural and overly formal. Readers react better to a human-sounding voice with simple language. Good rule of thumb: if you wouldn't say the word, don't write it.

Warning: Do not use long words to impress. Don't send your reader to the dictionary.

Do not use words ending in *-ize*: prioritize, finalize, actualize, finalize, conceptualize, utilize. . . .

Example:

Poor: Management initiated procedures for accident reduction.
Good: Management took steps to reduce accidents.
         (No short-word substitutes for *management* and *accident*.)

See also: Clarity—specific and concrete words.

Discussion:

Many long words creep in as cliches from the workplace. Use your dictionary and thesaurus to find short synonyms for long words.

You and the reader are stuck with some long words that have no good short word substitute: *civilization, dangerous, helicopter, satellite, computer.* . . .

You also use long words that take the place of many short words. For example, use *management* instead of *the people who watch over the firm's daily affairs.*

Exercise: Replace these long words with short, one-syllable words.
(Answer A-38)

1. accurate
2. actuate
3. additional
4. allocate
5. aggregate
6. apparent
7. ascertain
8. assimilate
9. assistance
10. capability
11. commence
12. constitutes
13. demonstrate
14. denominate
15. designate
16. disseminate
17. eliminate
18. enumerate
19. establish
20. expeditious
21. expertise
22. facilitate
23. functionality
24. generate
25. identical

26. initiate
27. magnitude
28. methodology
29. minimum
30. modification
31. necessitate
32. objective
33. operate
34. optimum
35. preliminary
36. prioritize
37. probability
38. remuneration
39. represents
40. self-conscious
41. sensible
42. stratagem
43. substantiate
44. suitable
45. terminate
46. uncompromising
47. underutilize
48. utilize
49. variance
50. voluminous

Technique 11.3   Break long sentences.

Tips:

Try breaking at punctuation marks—they often signal shifts in thought.

Break long series into vertical lists.

Keep your average sentence length less than 20 words.

Warning:

Do not break all sentences. Use long sentences carefully to express complex ideas.

Example:

Poor:

AAP selected BAP Inc. to design and develop critical applications for its internal management information system, which requires BAP to manage the construction of a data center, purchase all the necessary software, plan and implement the transition to the new system, integrate all existing records, operate and maintain the new system, then hire and train the new data center personnel.

Good:

AAP selected BAP Inc. to design and develop critical applications for its internal management information system. Therefore, BAP must manage the construction of a data center and purchase all the necessary software. Then BAP must plan and implement the transition to the new system, integrating all existing records. BAP must operate and maintain the new system, then hire and train the new data center personnel.

See also:

Economy; Coherence—vertical lists.

Discussion:

Recall editing for economy. If you cut your text by 20%, you cut your average sentence length by 20%.

The average spoken sentence length is 20 words. Professional writers keep average sentence length to less than 17 words.

Vary your sentence length with a range of 6 to 28-word sentences on each page.

Emphasize key ideas in short sentences, as short as six words. Develop complex thoughts with long sentences. Use vertical lists to group logically related items, statements, commands, or questions.

Exercise: Break these long sentences to improve readability. Use short sentences for emphasis, vertical lists to group related items, and long sentences to express complex relationships. (Answer A-39)

1. The client had told us that the tanker was purchased in December, 1986, and after fulfilling an existing obligation to act as a storage facility for fuel in the Caribbean, the tanker proceeded to Portugal in May, 1987, where it was dry-docked for barnacle scraping, painting, and repairs.

2. Because the multi-state Rentacar discount is the only discount plan that would require these types of functionality, and no other discount plans have been proposed that might require this functionality, the multiple levels or alternate level credits will not be addressed in any other functional specifications.

3. To meet the Air Force Controller's need for a financial system that would provide a single, consolidated repository of a budget execution, general ledger, and external reporting for Air Force-wide financial management purposes, we developed MegaCount software modifications; developed custom interface programs to provide the MegaCount application software with data from external Air Force budget execution and reporting applications; developed conversion programs to convert existing Air Force data to MegaCount formats and data files; and developed additional custom reports, including external reports for submission to Treasury and GAO.

4. The purpose of the Uniformed Securities Act is to protect investors from fraudulent securities transactions, for which the administrating agency requires securities to be registered with the state; and unless a security is specifically exempt from registration, or the transaction is considered exempt, the security must be registered before it can be sold, or offered for sale within the state.

Final exercise: The following passage has 40% long words and an average sentence length of 48 words. Cut deadwood, replace long words with short words, and break long sentences. Cut the long words to 10% or fewer, and reduce the average sentence length to less than 20 words. (Answer A-40)

Uninterruptible Power Supply (UPS) and Personal Computer Preventive Maintenance

Obviously, the configuration manager has an obligation to provide his or her corporation an electronically secure environment for the corporation's personal computers, and this dissertation will demonstrate an essential preventive maintenance application that provides reduction in the estimated mean time to failure rate for CPUs, improvements in hard disc performance, and increased reliability of input/output peripheral devices. The pre-eminent prerequisite for establishing a secure preventive maintenance environment is an excellent uninterruptible power supply (UPS), and we recommend the Microman Standby System with its internal EMI/RFI filters and surge protection capability, which provides the most economical protection for your equipment and your data from all serious electrical power interruptions. These marvelous uninterruptible power supplies (UPS) feature technologically superior power transference rates, a comprehensive set of diagnostic and LED status indicators, intelligent communication interfaces, audible alarms, and attractive casings, virtually eliminating power-related risk to sensitive electronic equipment.

Step 12. Check for Correctness

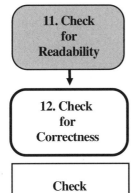

With your editing completed, check for correctness in the following order:

12.1   Check word choice.

12.2   Check grammar.

12.3   Check punctuation.

12.4   Check mechanics.

Discussion:

We use standard word choice, grammar, punctuation, and mechanics to communicate quickly and clearly. Departures from standards confuse and distract the reader.

Word choice refers to your use of individual words. **Grammar** is how you put the words together to express your thoughts. **Punctuation** marks indicate pauses and inflections. **Mechanics** are visual aspects, including layout and typography.

In his book *In Search of Excellence*, researcher Tom Peters notes that Delta Airlines meticulously cleans the coffee rings from the tray-tables, because some passengers assume that dirty tray-tables mean poor engine maintenance. Mistakes in word choice, grammar, punctuation, and mechanics are like those coffee rings: The reader may dismiss your message because of errors in correctness.

Technique 12.1 Check word choice.

Tips: Consult a good dictionary to check meaning and spelling.
Keep track of your common mistakes by putting a dot in the
margin of your dictionary when you look up a word.
Learn the meaning and pronunciation of words.

Check sound-alike words.

Warning: Do not trust your ears: Many word choice errors occur in
speech, then creep into our writing.

Do not rely on spell-checking software. The software doesn't
know if the word is misused, only if it's misspelled. Grammar-
checking software also falls short. For example, *"We can't
stand any more weeping and whaling"* would pass both spell-
checking and grammar-checking software.

Example:

Poor: *"Presently*, NIMROD does not operate on DOS machines."
Good: *"At present*, NIMROD does not operate on DOS machines."
(To learn the difference, look up *presently* in the dictionary.)

Commonly confused sound-alike words include:

affect, effect	assure, ensure, insure
council, counsel	compliment, complement
discreet, discrete	everyone, every one

See also: Readability; Correctness—mechanics.

Discussion:

English has more than 500,000 root words (some claim more than 1.2 million) with many
multiple meanings. Don't feel inadequate if you fail to memorize them all.

Use your dictionary to build vocabulary. First, own your dictionary—a good one like *The
American Heritage Dictionary*. Second, each time you look up a word, put a dot in the margin. When you find a dozen or so dots in the margin, you might as well memorize the word.
This dictionary-and-dot method ensures that you build your vocabulary with words you and
your colleagues actually use, instead of the lofty and often useless words memorized from vo-
cabulary-building tapes.

English has many sound-alike words, spelled differently, with different meanings. Writers
and readers often confuse these sound-alike words. You avoid sound-alike word confusion
(and improve readability) by using shorter words. For example, "Margaret (*instigates, initi-
ates*) a (*discrete, discreet*) set of commands to put the computer to (*continuous, continual*) operation" confuses the reader. "Margaret *begins* a *distinct* set of commands to *repeat* the
computer operation" communicates clearly, simply, and correctly.

Exercise: Define these commonly confused words. Circle definitions that match the italicized word, then write sound-alikes that match the remaining definitions. (Answer A-41)

Confused words	Definitions	Sound-alikes
1. *except*	to receive with favor (aside from)	*accept*
2. *adapt*	highly skilled to take as one's own to adjust to the situation	
3. *advice*	a noun meaning counsel given a verb meaning to recommend	
4. *effect*	to produce a change a result (noun) to result in (verb)	
5. *ensure*	to promise someone to make sure to protect against loss	
6. *capital*	seat of government money owned the building where legislators meet	
7. *cite*	to use as proof to summon to appear in court act of seeing that which is seen place or location	
8. *compliment*	that which completes a flattering comment	
9. *counsel*	a group of people advice (noun) advise (verb)	
10. *discrete*	tactful separate or distinct	
11. *everyone*	every person of a group every person	
12. *farther*	space or distance to a greater degree	

Confused words	Definitions	Sound-alikes
13. *formally*	according to custom in the past	
14. *it's*	a possessive pronoun contraction of it is	
15. *lie*	to place to recline	
16. *lone*	the act of lending (verb) that which is lent (noun) by oneself isolated	
17. *lose*	not fastened or confined to part with	
18. *past*	moved on (verb) at a former time (adjective) former time (noun)	
19. *personal*	private employees	
20. *principal*	leader money (noun) first or highest (adjective) rule	
21. *respectively*	showing respect considered singly	
22. *write*	upright correct solemn act to make words on a surface	
23. *stationary*	standing still letter paper	
24. *statue*	height or level rule or law carved figure	
25. *than*	in comparison with at that time	

Confused words	Definitions	Sound-alikes
26. *their*	possessive pronoun at that place contraction of "they are"	
27. *too*	toward in addition one more than one	
28. *addition*	increase attachment publication	
29. *alter*	religious table change	
30. *basis*	reasons, or foundations a reason, or a foundation	
31. *biennial*	twice a year every two years	
32. *devise*	equipment plan	
33. *disapprove*	have an unfavorable opinion show to be false	
34. *elicit*	ask for illegal	
35. *illegible*	qualified unreadable	
36. *eminent*	prominent about to happen	
37. *envelop*	surround container for a letter	
38. *expend*	increase pay out	
39. *physical*	financial of material things	
40. *forward*	preface at the front	

Technique 12.2 Check grammar.

Tip 1: Check sentence structure.
Be sure

- each sentence has at least one subject and verb
- each sentence expresses at least one complete thought
- thoughts join together with standard punctuation

Warning: Do not mistake incomplete thoughts for complete thoughts. Connecting words such as *after, when, because,* and *while* make thoughts incomplete.

Incomplete: They postponed the survey. *Until we finish the fall budget.*
Complete: They postponed the survey until we finish the fall budget.

Example:

Each sentence must have at least one subject and verb.
Missing verb: Carol Adams, newly elected president of the Jaycees.
Complete: Carol Adams, newly elected president of the Jaycees, *entered the room to loud applause.*

Each sentence must express at least one complete thought.
Incomplete: Whenever John arrives late to work.
Complete: Whenever John arrives late to work, *he blames traffic.*

Thoughts must join together with standard punctuation.
Run-on: Success has a thousand fathers failure is an orphan.
Corrected: Success has a thousand fathers; failure is an orphan.

Comma Splice: A word to the wise is sufficient, don't drink and drive.
Corrected: A word to the wise is sufficient: don't drink and drive.

See also: Clarity—parallelism; Correctness—punctuation.

Discussion:

Check for these common mistakes in sentence structure. For a more detailed grammar review, use grammar handbooks like the *Harbrace College Handbook* available at your local bookstore.

A run-on is two or more complete sentences joined without punctuation.

A comma splice occurs when a comma joins two or more complete thoughts.

Exercise: Circle the subject(s) and underline the verb(s) of each sentence. (Answer A-43)

1. Thunder is loud, but lightning does all the work.

2. The customer does not know what we can do for her company.

3. Fred and Barney took Wilma and Betty dancing.

4. Your contribution to the project deserves our praise.

5. Sticks and stones may break my bones, but words will never hurt me.

6. You can't win if you don't play.

Exercise: Identify each word group below as a correct sentence (C), incomplete thought (IT), run-on (RO), or comma splice (CS). (Answer A-44)

1. Considering that the competition has reacted strongly to our effort to grab more market share.

2. Can type twenty-five words per minute.

3. Mr. Johnson, unable to attend the afternoon meeting or evening dinner.

4. Beverly Timmons, project leader for database development, made three unsuccessful requests for government assistance.

5. Whose responsibility is it to clean up the oil spill?

6. Until we found out that Good Food Inc. had raised its price to cater a cocktail party and the Sheraton Inn had almost doubled the price to rent the ballroom.

7. Now that Sandra has finished her Associate Degree in Accounting.

8. Tax increases choking off economic growth again.

9. The office manager interviews all candidates for staff positions.

10. Friendly, courteous, and always available to answer your questions about our software products.

11. The favor of reply is requested.

12. We have a scheduling conflict for the conference room, Mr. Smith scheduled a news conference at four and the facility engineer planned to recarpet the floor, please advise.

13. The climb to the top is hard remember that staying at the top is harder.

14. Concentration is the key to economic success, it's also the key to success in life.

Step 12. Check for Correctness

Tip 2:   Check pronouns.
              Be sure

- subject and object pronouns use correct form
- possessive pronouns use standard spelling
- collective nouns and pronouns agree in number

Warning: Do not confuse possessive pronouns with contractions.

pronoun:	your	its	whose	their
contraction:	you're	it's	who's	they're

Example:

Use subject and object pronouns to substitute for a subjective
or objective noun. "*We* (subject) found *them* (object) in the corridor."

Most possessive pronouns do not use an apostrophe.
Possessive pronouns include *mine, his, hers, ours, yours, theirs, whose.*

Use a singular pronoun when a collective pronoun acts as a
group. "The *jury* reached *its* verdict in less than two hours."

Use a plural pronoun when the members of a collective noun
act separately. "The *jury* left for *their* homes at day's end."

Discussion:

Recall editing for clarity. You've already eliminated ambiguous pronouns, thus solving many
if not most pronoun problems.

Pronouns	*Subjects*	*Objects*
1st person	I, we	me, us
2nd person	you	you
3rd person	he, she, they, it	her, him, them

Use these techniques to choose the subjective or objective pronoun that will substitute for a
subjective or objective noun.

1. Reverse a sentence that has a "to be" verb to choose the right pronoun. For example, *That's
him* is obviously incorrect when reversed to *Him is that.*

2. Be careful when using two or more pronouns together, or when using a pronoun and a
noun. Choose the right pronoun by imagining one of the words left out. For example, *Give it
to Mary and I* is obviously incorrect when reduced to *Give it to I.*

3. Complete the sentence with understood words to help you choose the correct pronoun. Use
this method when the sentence includes the words *than* or *as.*

No one loves you more than your father and (I, me).
Complete the sentence and it reads "than your father and I *do.*"

4. Sometimes the pronoun you choose determines the meaning of the sentence. Be sure you
communicate the intended meaning.

Do you call the office more often than (I, me)?
Do you call the office more often than I (*do*)?
Do you call the office more often than (*you call*) me?

Exercise: Circle the correct pronoun. (Answer A-45)

1. The two winners were Jane Swanson and (I, me).

2. Please send Mr. Jenkins and (I, me) to the seminar.

3. Both you and (he, him) should apply for the new position.

4. The telephone technician (who, whom) you sent for has helped us before.

5. Rebecca is taller than (I, me).

6. No one wants to win the AIMS job more than Alice Cairns and (me, I).

7. The company must monitor (its, their) sick leave policy carefully.

8. BAP Industries, Inc. has (its, their) headquarters in Virginia.

9. The team won (its, it's, their) first game of the season.

10. (Its, It's) not (I, me) (whose, who's, who am) responsible for losing the key!

11. (Their, They're) talking about (your, you're) book.

12. (Whose, Who's) in charge of marketing?

13. The committee can't agree what (its, it's, their) responsibilities are.

14. That's (he, him) standing in the lobby.

15. Jerry writes better than (they, them), so (their, they're, there) supervisor asked (he, him) to edit the company newsletter.

Tip 3:   Check verbs.
Be sure

- subjects and verbs agree in number
- special verb form *were* expresses unreal ideas

Warning: Do not assume the noun closest to the verb is the sentence's subject. Note the italicized subjects and verbs below:

IBM, like other computer manufacturers, *has* lowered prices.
Each of the three proposed plans *ensures* good results.
Skilled *masters* of ceremony *don't start* a speech with a joke.

Example:

A singular subject uses a singular verb; a plural subject uses a plural verb.
  A *collection* of paintings by the local artists *is* on display .
  Our short-term *goals* and long-term *budget* conflict.

A collective noun acting as a group uses a singular verb.
  The *faculty uses* the conference room for its meetings.

When its members act separately, a collective noun uses a plural verb.
  The *faculty wear* their caps and gowns at graduation.

Use the verb form *were* to express unreal ideas (subjunctive mood) regardless of the number of the subject.
  If *he were* more ambitious, John could become president.

See also: Clarity; Economy.

Discussion:
Recall editing for clarity and economy. You've already made your prose clearer by using present tense, active voice, and imperative or indicative mood verbs. In addition, you cut deadwood associated with buried verbs.

Exercise: Circle the subject. Write the verb in the form that agrees with the subject. Some sentences are correct. (Answer A-46)

1. Each of the four divisions in the company is responsible for its own costs.

2. Neither the Army nor the Air Force wants to pull troops out of Germany.

3. The senior scientist and engineer in this company wants to work on the space-station contract.

4. Difficult decisions like the one we must make today takes time.

5. A collection of paintings by three local artists is on display in the lobby.

6. A four-member crew cleans and maintains each UPS truck.

7. Where does the desk, chair, sofa, and filing cabinet go?

8. Shoes, belt, and a tie add a lot to a man's wardrobe.

9. The board of directors agree with management.

10. The carton of typewriter ribbons are sitting on the desk.

11. The duties of the police officer requires courage and self-sacrifice.

12. Attention to details ensure fewer errors.

13. Both have the authority to write checks up to $1,000.

14. George Burns, with his companion Gracie Allen, needs no introduction.

15. Here is the new copy machine and its instruction manual.

16. Half a load of bricks do not satisfy our order.

17. I wish I was your boss instead of your assistant.

18. If wishes were horses, then poor men would ride.

19. If I was a full time employee, I would get a salary with benefits, but I would lose my overtime.

20. ACME Theaters is a large national chain.

Tip 4:   Check adjectives and adverbs.
Be sure

- *a*'s and *an*'s agree with the sound of the word that follows

- only one negative word is used in a negative statement

Warning: Do not decide to use *a* or *an* simply on the appearance of the first letter of the next word. Letters can be pronounced in different ways.

an umbrella   a unit   an hour   a hope

Example:

Use the word *an* if the following word begins with a vowel *sound* (a, e, i, o, u). Otherwise, use *a*.

an angle	an egg	an intern	an option
a one-day job	an umpire	a utility	

Use only one negative word in a negative statement: *He hasn't received any orders.* A double negative results when you use two negative words in a negative statement: *He hasn't received no orders.* Logically, a double negative equals a positive—not your intended meaning:
We have not scarcely begun means *We have earnestly begun.*
The plane barely missed. . . means *The plane directly hit . . .*

A sentence with more than one thought can use a negative word in each thought. (If you *don't submit* your application, you *won't be considered* for the job.) However, when editing for clarity, make your sentences positive. (If you *submit* your application, you *will be considered* for the job.)

See also: Clarity—be positive; Economy.

Discussion:

Recall editing for clarity and economy. You've already worked to use concrete and specific terms, eliminating many adjectives and adverbs. Then you cut unnecessary adverbs and adjectives as part of cutting deadwood.

Negative words include *no, not, neither, never, none, nowhere, nobody, scarcely, rarely, barely, seldom, don't, doesn't, won't, can't, shouldn't, aren't, wouldn't, couldn't, haven't, hardly.*

Exercise: Write *a* or *an* before each word. (Answer A-47)

1.___10 percent raise
2.___action
3.___example
4.___European
5.___hour
6.___hostess
7.___MBA
8.___order
9.___uncle

10.___11 percent drop
11.___balancing act
12.___donor
13.___FBI investigation
14.___history book
15.___icon
16.___one-time write off
17.___S.O.S.
18.___uniform

Exercise: Write *a* or *an* in each blank. (Answer A-47)

1. _____ one-month night shift is followed by _____ eleven-day paid leave.

2. Dr. Peters, ___history professor, taught ___ unit about the Civil War.

3. __ ambassador from __ European country made __ unusual request at __ UN meeting.

4. His clock has __ electrical dial, not __ hour, __ minute, or __ second hand.

5. _____ aspirin is not always enough for _____ aching head.

6. Jane earned _____ M.B.A. with _____ emphasis on marketing.

7. We need 100 days to complete ___ order, but we have ___ 83-day deadline.

8. The doctor told me to start ___ aerobic activity, which was not ___ answer I wanted to hear.

Exercise: Correct the double negatives in these sentences. (Answer A-48)

1. James couldn't find hardly anyone to invest in his gourmet doughnut shop.

2. Nobody doesn't want to miss the staff meeting.

3. Peter doesn't know nothing about the value of a dollar.

4. If you don't have a positive attitude, you won't succeed.

5. She couldn't hardly hope to get a promotion after one week on the job.

6. Wouldn't Jim rather not go?

7. Let's not give no more thought to the unfortunate incident.

8. Phyllis never hardly saw no records to justify Bill's tax deductions.

9. Roger never met nobody that didn't not like his mom's toll house cookies.

10. The safety inspector told us that we must not store none of the nitro next to the glycerin.

Technique 12.3 Check punctuation.

Tip 1:	Check apostrophes. Be sure

- apostrophes show possession correctly
- apostrophes show plural correctly

Warning: Do not confuse possessives with plurals.

Do not add an apostrophe to the plural of capital-letter abbreviations (2 IBMs, 3 RFPs).

Example:

Add *'s* to a make a singular noun possessive.
> The *sailor's* boat sank.
> Exception: If a singular noun ends in *s* and has more than one syllable, add only the apostrophe.
> Mr. *Childress'* hat blew off.

Add an *'* to make a plural noun that ends in *s* possessive.
> *The soldiers'* relatives were waiting for news.
> The *Joneses'* home is beautiful. (more than one person named Jones)

Add *'s* to make a plural noun possessive if it does not end in *s*.
> *Men's* suits can be expensive.
> Your future *sisters-in-law's* dresses arrived two weeks before the wedding.

Use an apostrophe for the plural of letters, words, and numbers.
> How many *t's* did you count?
> Watch the use of *and's* in your writing.
> Cross out the *2's*.

See also: Correctness—mechanics.

Discussion:
An apostrophe and *s* are often used to indicate possession.
Possessive relationships include
- personal—John's friend
- ownership—the women's shoes
- time—a year's leave
- authorship—the accountant's report
- type or kind—men's clothes

Exercise: Write the singular possessive and the plural possessive. For example, *player* has singular possessive *player's* and plural possessive *players'*. (Answer A-48)

1. clerk	5. Jones	9. day	13. business	17. man
2. facility	6. area	10. boss	14. knife	18. advisor
3. fence	7. city	11. guest	15. line of credit	19. loss
4. waitress	8. year	12. lunch	16. brother-in-law	20. friend

Exercise: Insert an apostrophe and an *s* to show a possessive noun. Make other nouns plural if necessary. (Answer A-49)

1. The employee attend lectures where they learn techniques to improve their plant efficiency.

2. We investigated Bill complaint.

3. All of the question were answered in turn.

4. The Treasurer recommendation is that we cut overhead cost.

5. If we build a men locker room, we'd better build a women locker room too.

6. We hired several new employee for the Jason project.

7. Fred office will need two coat of paint.

8. It's the office manager responsibility to make sure the light work in the conference room.

9. The toxic waste response team must respond in a minute notice.

10. Please evaluate Avis proposal to discount our company large car rental fees.

Tip 2:	Check commas. Be sure

- commas separate items in series
- commas separate introductory expressions from main idea
- commas set off parenthetical expressions

Warning:	Do not forget the second comma in a pair of commas.
Poor:	Senator Foghorn, our senator runs for re-election this year.
Good:	Senator Foghorn, our senator, runs for re-election this year.

Example:

Use commas to separate items in series, including before the *and*.
> Betty wrote, edited, proofread, and produced the manual.

Use a comma to separate most introductory expressions from the main idea of the sentence. Use commas after introductory expressions that contain any form of a verb.
> *If she calls*, tell her I'm out.
> *Planning ahead*, Marty packed his tennis shoes.

Use a comma after an introductory expression of five or more words.
> *In the back of the computer*, the serial port connects to the printer cable.

When an introductory expression has fewer than five words and has no verb form, you decide if a comma is needed.
> *In our county, taxes* are high. (comma needed to distinguish *county* from *county taxes*)

Use commas before and after a non-essential expression—details that add to a sentence but do not change its meaning.
> The result, *however,* was disastrous.
> Thomas, *a chemical engineer,* took the night shift.
> Refer to the letter, *which I sent last week,* for your answer.

See also:	Clarity.

Discussion:

Commas indicate a short pause to separate items. To help yourself place commas correctly, consider where you pause briefly in speech.

Use *which* when the expression is non-essential, and set off with commas. You can remove non-essential expressions without changing the sentence's meaning. Use *that* for an essential expression and do not set off with commas.

Read my letter, *which* I sent March 11. (Non-essential—I sent you that one letter only.)
Read my letter *that* I sent March 11. (Essential—I have sent you other letters.)

Exercise: Insert commas where needed. (Answer A-49)

1. The Smith Foundry Tool and Die Company was started by Thomas Smith an inventor and entrepreneur.

2. I want to see last quarter's income statement balance sheet and cash flow statement.

3. Margery Elizabeth Susan and I can just barely fit in her new sedan.

4. IBM Compaq Apple and a host of other personal computer manufacturers are struggling to define their marketing strategies.

5. Employee benefits include paid vacation holidays sick leave bereavement leave and unpaid maternity leave.

6. The new copy machine is faster cleaner more reliable and more versatile.

7. We can buy industrial grade fire retardant carpet in either beige dark blue green gray or burnt orange.

8. Our lounge always keeps pots of regular and decaffinated coffee with cream and sugar.

9. Your flight has stops in Atlanta Denver Los Angeles and Melbourne.

10. Because the air conditioner broke down we're releasing the workers at 2:00 p.m.

11. After I just got a $2 million construction loan you've got a lot of nerve telling me you underestimated the job.

12. Although they finished paving the parking lot they have not painted the lines yet.

13. Gigamega the most powerful computer ever built has been programmed to invent video games.

14. Our new vice-president for engineering Dr. Potts will lead the discussion on cryogenics.

15. William's plan even though it made no sense to us won high praise from the Navy.

16. If I had to learn a second language all things being equal I would study FORTRAN.

17. Patriots Day a paid holiday in Massachusetts does not merit a day off in Virginia.

18. Slick Magazine boasting a circulation of 5 million paid subscribers charges $1600 for a quarter-page ad.

19. In the past success came easily for George.

20. In conclusion we use commas to separate parenthetical expressions from the main idea of the sentence.

Tip 3:   Check semicolons.
Be sure

- semicolons join two complete thoughts correctly
- semicolons separate list items that have internal commas

Warning: Do not use semicolons to join a complete thought with an incomplete thought or phrase.

Poor: Although Inga lives in Maryland; she works in Virginia.
Good: Although Inga lives in Maryland, she works in Virginia.

Poor: The project manager met with the client; because the contract requires weekly reviews of schedules, costs, and deliverables.
Good: The project manager met with the client, because the contract requires weekly reviews of schedules, costs, and deliverables.

Example:

Use a semicolon to join two closely related, complete thoughts.
Ms. Kim's references were excellent; we offered her the job.
Bob liked the car; he could not afford it.

Note: Transition words such as *therefore, however, and nevertheless* often relate ideas. Use a semicolon before these transitional expressions to join complete thoughts. Use a comma after the transition word.
Bob liked the car*; however,* he could not afford it.

Use semicolons to separate items in a series when the items already have commas.
They visited *Atlanta, Georgia; Tampa, Florida;* and *Mobile, Alabama,* during their spring sales trip.

See also: Clarity—parallelism.

Discussion:

The semicolon marks a pause longer than a comma, but shorter than a period (notice that it consists of a period over a comma.) A semicolon often takes the place of the word *and*.

Exercise: Insert semicolons where needed. (Answer A-50)

1. The receptionist area needs new carpet, however, we'll wait until we remodel the entire floor.

2. Although the receptionist area needs new carpet, we'll wait until we remodel the entire floor.

3. Dr. Latrobe, a propulsion expert, designed a rocket motor that runs on normal jet fuel, nonetheless, liquid hydrogen remains our preferred fuel, because it has a better thrust to weight ratio.

4. Francis will meet us at O'Hare Airport, however, thirty minutes later than expected.

5. Luck is where preparation meets opportunity, so keep your eyes open and be prepared.

6. We won the contract, now we have to do the work.

7. Karen Kelly brought us some of our most profitable accounts, for example, she landed both the Hechinger and the Safeway accounts.

8. We've added four new sales districts, which are Atlanta, Georgia, Mobile, Alabama, New Orleans, Louisiana, and Houston, Texas.

9. Our company has but one mission, that is, to provide our clients the best value in video home entertainment.

10. Megatech bid the highest price, nevertheless, they won the contract on technical merit.

11. Conglomerator, Inc.'s most recent acquisitions were Catfish Farms, Ltd. on April 10, 1989, and Carlisle Cosmetics, Inc. on December 2, 1990.

12. Although Mr. Derickson is younger than the other applicants, he should get the job because of his superior performance record.

13. Our company's policy is to promote from within, for instance, Mr. Jacobs started as a clerk and rose to be chief executive officer.

14. Because John Heath just came to us from the Department of Transportation, where he was a special assistant to the Secretary, we mustn't bid him on the Highway Study Project.

15. As long as sales continue to increase at the present rate, we can absorb the rising cost of labor without raising our prices.

Tip 4:	Check colons. Be sure

- colons come after a complete thought when a supporting thought, phrase, or word follows
- colons come after a complete thought when a list follows

Warning:	Do not use colons after an incomplete thought.
Poor:	After the blades have stopped completely: remove the cover to inspect the roller bearings.
Good:	After the blades have stopped completely, remove the cover to inspect the roller bearings.

Poor:	The office exercise room has: free weights, stationary bikes, and a sauna.
Good:	The office exercise room has free weights, stationary bikes, and a sauna.

Example:

Use a colon after a complete thought when a supporting thought, phrase, or word follows. When a complete thought follows a colon, capitalize it.

> *Tom Sawyer is my favorite character: He never gives up.*
> *His answer was simple: no.*

Use a colon when a complete sentence introduces a quotation.

> *The president made this point: "We must all pull together."*

Use a colon after a complete sentence when a list follows.

> *We suggest these furnishings: a desk, chair, table, and lamp.*

See also:	Clarity; Correctness—grammar.

Discussion:

Colons indicate that supporting details follow. Use colons only after a complete thought, never after an incomplete thought.

Use colons after a complete thought that introduces a vertical list.

Exercise: Insert colons where needed. Check capitalization. (Answer A-51)

1. Next time, give that pushy salesman an evasive answer, tell him to take a long walk off a short pier!

2. Note well the company would have posted a substantial loss in 1990 except for the one-time sale of the Occoquan property.

3. Our company has but one mission to provide our clients the best value in video home entertainment.

4. Conglomerator, Inc. has made two acquisitions Catfish Farms, Ltd. on April 10, 1989; and Carlisle Cosmetics, Inc. on December 2, 1990.

5. You must add one procedure to lower your worker's compensation insurance you must aggressively prosecute fraud.

6. Eric made a significant breakthrough in his research he discovered a new graphite compound.

7. Doc Watson put a sign on his briefcase Moon or Bust!

8. The odor from the paper mill smoke smells bad, but it's harmless.

9. Managing inter-personal conflict is like the law of thermal dynamics you can't win, you can't break even, and you can't get out of the game.

10. The employee lounge has three simple rules for everyone's mutual enjoyment no smoking, no alcoholic beverages, no radio-players without earphones.

Tip 5:   Check dashes.
        Be sure dashes emphasize details or thoughts.

Warning: Do not confuse a hyphen with a dash. A dash is as wide as a capital letter 'M' or twice as wide as a hyphen. If you can't print an M-dash or N-dash—so called because the dash length equals the width of capital M or N—make your dash with two hyphens, with no space on either side.

        Do not overuse dashes. When everything screams for attention, nothing gets noticed. Too many dashes fragment and disorganize thoughts.

Example:

Use dashes to emphasize a non-essential expression that you usually enclose with commas. For less emphasis, use parentheses.

> *Our firm, which earned a large profit last year, hopes to expand into new markets.*

> *Our firm—which earned a large profit last year—hopes to expand into new markets.* (emphasized)

> *Our firm (which earned a large profit last year) hopes to expand into new markets.* (de-emphasized)

Use dashes to separate a nonessential expression that contains one or more commas. To de-emphasize the word group, use parentheses.

> *The neighbors—Mr. Johnson, Mr. Black, and Mrs. Smith—attended the party.*

Use a dash after a series or a single word that comes before a complete thought.

> *Looks, brains, talent*—she had them all.

See also: Correctness—hyphens.

Discussion:

For expressions within a sentence, use

- dashes to emphasize the expression
- parentheses to de-emphasize the expression
- commas to give neutral weight to the expression

Use parentheses—not dashes—to enclose directions. *The figures (see page 2) indicate modest gains in 1991.*

Exercise: Insert dashes, parentheses, commas, or colons. Some sentences can be punctuated several ways. (Answer A-51)

1. The invoice for $215.00 not $21.50 needs your prompt attention.

2. Writing and editing ability that's what we want in our senior technical staff.

3. The partnership usually pays a portion of the net profits for example, $670 per limited partner in 1990 to help cover the limited partners' tax liability.

4. Jeff must fix the rear projector in the conference room today tomorrow is too late.

5. The Penultimate II cordless phone you won't find a better value offers the following features speed dialing, auto call back, conference calling, and much more.

6. The offices on the the sixth floor Treasury, Marketing, and Human Resources will be moved to the new building in May.

7. Dorothy had it all a dog, ruby slippers and Kansas.

8. Rebecca requested a four-week vacation she won a cruise and she has no choice as to dates.

9. Sales rose see figure 4, page 32 to a record high however return on sales fell slightly.

10. A high school diploma, three years' experience, good references these are the minimum requirements.

Tip 6:   Check hyphens.
                    Be sure

- hyphens separate parts of a word
- hyphens join parts of a compound adjective

Warning: Do not confuse a hyphen with a dash.

Example:

Use a hyphen to separate parts of a word.
twenty-two
self-taught

Use a hyphen to join the parts of a compound adjective (an adjective made up of two or more words). In a compound adjective, the first adjective describes the second adjective. For example,
a *well-oiled* machine (*well* describes *oiled*)
a *computer-generated* graphic (*computer* describes *generated*)

Use a hyphen when a compound adjective appears before the noun.
Do not use a hyphen when the phrase appears after the noun and verb, or if the first word in the phrase ends with *ly, er,* or *est.*
Sue produced a *well-written* memo.
The memo was *well written.*
The *slowly opening* door caught his attention.

See also: Correctness—dash.

Discussion:
Use a hyphen in other compound words as indicated by your dictionary or company style manual. For example, *data-base* or *database* might be standard.

Exercise: Punctuate compound adjectives with hyphens. (Answer A-52)

1. We graduate a hundred odd students each year.

2. John sent a carefully worded letter to IRS to explain his highly irregular filings for 1988 and 1989.

3. Send an up to date roster of security clearances to Lt. Avery.

4. The security inspector, Lt. Avery, said our security clearance roster was not up to date.

5. It's a pleasure to receive a well written proposal.

6. He wished his off the record remarks had stayed off the record.

7. Alice made a reasonably good attempt to send the package before five o'clock.

8. John and Martha sublet a one bedroom apartment in a not so nice part of town.

9. Thelma told the interior decorator she wanted egg shell white paint in the dining facility, but when the paint dried, she swore the color was coffee stain brown.

10. Mrs. Stern won't tolerate a gum chewing receptionist.

Tip 7:   Check vertical lists.
Use standards for punctuating vertical lists. Consult the standards established by your style guide, or use these American business standards to punctuate your lists.
Be consistent.

Use a colon after an introductory *sentence.*
Omit the colon after an introductory *phrase.*

If list items are complete *sentences*, begin with a capital and punctuate them as complete sentences with ending period, question mark, or exclamation mark.
If list items are *not complete sentences*, omit ending punctuation.

Warning: Do not vary the way you punctuate lists.

Example:

Pert Plan software offers you three options for listing a task:
      1. Give task with a firm deadline.
      2. Give task an estimated range of time for completion.
      3. Give task an average time to completion.

The project start-up team must

- provide a work plan
- negotiate definitions for deliverables
- hire and train project staff

See also: Clarity—parallelism; Correctness—colons.

Discussion:
Make each item in your list grammatically parallel; make punctuation at the end of each item consistent.

Modern business and technical writers favor using less punctuation in vertical lists for a cleaner, more streamlined style. Punctuating a vertical list with ending semicolons and period looks old-fashioned.

Exercise: Correct the punctuation in these lists. (Answer A-52)

1. We need to address two conversion issues for Design Release 2.2. These are:

    1) the conversion of information from the VAX to IBM environment; and,
    2) the initialization and maintenance of the operator's manual.

2. Avoid ambiguity:

    1) choose words carefully
    2) place modifiers close to things they modify
    3) use active verbs and avoid passive voice

3. Before you turn off the LAN host computer

    Close any open files and exit any active programs
    Run the backup to tape procedure
    Run the LAN check to warn users of system shutdown
    You must input a valid LAN operator ID# then end the LAN program

4. The strengths of the Shazbot system include:

    a. error trapping prevents faulty data entry

    b. on-line help functions decrease training time

Technique 12.4 Check mechanics.

Tips:	Be consistent in layout and typography. Follow or establish visual standards. Use or create a style guide.

Warning: Do not vary mechanics without a logical reason.

Example:

Poor:

Memo to: Karen Black
From: Bill Martinson
Date: May 21 1994
Subject: '94-95 Project control weekly Status report,
             WE 5/21/94.

Good:

To: Karen Black
From: Bill Martinson
Date: May 21, 1994
Subject: 1994-95 Project Control Weekly Status Report
           Week Ending May 21, 1994

See also: Coherence—visual devices.

Discussion:

Mechanics are the visual aspects of a document, including layout and typography. Layout refers to the arrangement of type on the page. Typography refers to the use of the type itself. Examples of layout and typography include

Layout	**Typography**
margins	capitalization
indentation	fonts
pagination	acronyms
titles and headers	abbreviations
ragged right or justified	numbers
spacing	spelling

Remember, you used visual devices when editing for coherence to help the reader skim, group, and order information. Now check that your mechanics are standard and consistent to help the reader understand and follow your discussion easily. Unnecessary shifts in mechanics—like unnecessary shifts in key words—confuse and distract the reader.

Although standards for word choice, grammar, and punctuation remain fairly constant, standards for mechanics vary. Learn available standards. Use a style guide that determines mechanics for your profession, company, project, or document. Create your own mechanics style guide if none exists. Then follow your visual guidelines consistently.

Exercise: Improve the mechanics of these sentences. (Answer A-53)

1. At our 9:00 meeting we reviewed the 4-month extension through 9/90. We learned yesterday that DoD (Dept. of Defense) will request another two month extension. We decided to submit the extension to DOD for 2 months with two one month options.

2. Beginning 3/19/95 Ben Brown will handle any material and/or supplies request through the MS/Plus computer system.

3. The mayor introduced former pres. Bush at the local Veteran's day celebration in Alex., VA.

4. Enter your userid in the ibm; then transmit your lotus files from the pc's harddisk to your own floppy disk.

5. 250 people attended the air force convention in Palm springs, CA.

6. Mr. Smith called this morning. (he left his telephone and fax #)

7. The order was for 12 6 inch pipes; We shipped them to Joe's hardware inc. yesterday at 5 pm.

8. See figure four on page nine.

9. My favorite book is How to Repair Your Volkswagen: a step by step manual for the complete Idiot.

10. This occurrance only strengthens our comittment to proceding with BAP industries' expansion into nickle mining.

Step 12. Check for Correctness

Final exercise 1: Circle the correct word. (Answer A-53)

1. We are *(adapt, adept)* at software design.

2. Please indicate your *(ascent, assent)* by signing the contract.

3. The improved lighting has had a good *(affect, effect)* on productivity.

4. Careful pre-writing *(assures, ensures, insures)* effective writing.

5. Dewey, Cheetham and Howe serves as *(council, counsel)* to the city *(council, counsel)*.

6. The salad makes a fine *(compliment, complement)* to the broiled fish.

7. It helps to break the problem into *(discreet, discrete)* topics.

8. *(Everyone, Every one)* must attend the safety briefing.

9. We cannot discount our hourly rates any *(farther, further)*.

10. Can we *(forego, forgo)* the interview process?

11. Mr. Smith was *(formally, formerly)* self-employed.

12. The operator must get *(past, passed)* the shut-off valve before seeing the display.

13. We will not allow an employee administrative leave unless we believe there is a serious *(personnel, personal)* problem.

14. The key to persuasive writing is seeing the reader's *(perspective, prospective)*.

15. The students felt sure of success because they had a *(principal, principle)* at stake.

16. Please examine each cost item *(respectfully, respectively)*.

17. Please *(sit, set)* yourself a place at the table.

18. What harm is a couple of beers *(between, among)* friends?

19. The Fairfax County Symphony gave a *(credible, creditable)* performance.

20. Doctors recommend we eat a *(healthy, healthful)* breakfast.

21. There are *(less, fewer)* than six days left to complete the work.

22. We moved our offices to the suburbs *(since, because, due to)* the lease expired and *(since, because of, due to)* the high prices in the city.

23. As the company's founder, Mr. Adam Smith raised the company to a *(respectable, respectful)* position in the steel industry.

24. Refer to the letter, *(that, which)* I sent last Tuesday.

25. The lawyer had no *(further, farther)* questions for the witness.

Final exercise 2: Punctuate these sentences correctly. (Answer A-54)

1. If ever you've nothing to do and plenty of time to do it in why don't you come up and see me.—Mae West in the movie "My Little Chickadee."

2. In his best selling book <u>Wabbit Hunting</u> the author Elmer Fudd discusses a hundred ways to trap snare or shoot cwazy wabbits.

3. Our bookkeeper Teresa impressed the auditors with her accurate files.

4. The temporary services agency Temps & Co. will give us eight hours of temp services at no charge just so we can evaluate their company.

5. Erica the company expert on time management suggests that we conduct all staff meetings standing up.

6. We billed four hours at the principal rate of $180 per hour and eight hours at the staff rate of $42 per hour.

7. Lorna Ewald Ph.D. in computer sciences started her own company in 1985 but she sold her interest to Logicon Inc. and then she came to work for us.

8. Population growth in the United States according to the latest census data has fallen if you take out immigration.

9. The qualities we seek include good people skills willingness to learn and willingness to travel.

10. Sam's motto cash is king makes a lot of sense in the 90s when so many companies struggle under debt.

11. Steel, oil, and railroads the great monopolies of the 19th Century changed the face of capitalism forever.

12. Dr. Nathan our only nuclear engineer decided that the company's research into cold fusion is a poor investment.

13. A typical engine overhaul is a one day job.

14. Red, white and blue will wrap our Fourth of July Sale in the flag.

15. The telecommunications van must be able to operate in the tropics therefore we added a dehumidifier to its on board equipment.

16. Dr. Harold Brown Chairman of the Loudon Board of Trade met with the Loudon County Zoning Commission to attempt a compromise between local environmentalists and developers.

17. AMTRAK's Metroliner runs between Washington, D.C. and New York in 2 hours and 52 minutes with stops in New Carrolton Maryland Baltimore Maryland Wilmington Delaware and Philadelphia Pennsylvania.

18. Any member of our cross trained staff that would include me can help you solve your most difficult files management problems.

Step 12. Check for Correctness

Final exercise 3: Find and correct the word choice, grammar, punctuation, and mechanics errors in this excerpt. (Answer A-55)

Forecast Economics pride itself for the results their clients receives from there many time-tested econometric model. The success of a model, depends on the assumptions upon which they are based. To assure the accuracy of the assumptions it uses, several design are tested to find the sensitivities of shifts in the assumptions. Moreover, Forecast researched brokerage costs, price entry, and looked at exchange fees. It's research into specific market securities have also addressed, the risks of volatility, or any other risks that come to mind.

Completeness and accuracy are the goal. For example dr. Brown will incorporate into the model any new regulatory information or new market theory. Whenever the thought occurs to himself. These flexible and rapid response capability couldn't hardly be replicated anywheres else. In the past, he have had too determined and quantify the extent that the regulatory aftermath of the 10/19, 1987 stock market crash would be having on the standard and poor's Stock Index. Having successfully incorporated these kind of market aberration into econometric modeling, clients have relied on Forecast Economics for many of they're real-time data modeling challenge.

Step 13. Proofread

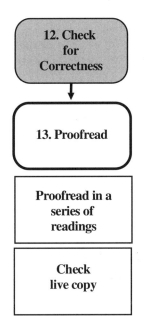

After correcting your document's word choice, grammar, punctuation, and mechanics, use two techniques to proofread accurately and quickly:

13.1   Proofread in a series of readings.

13.2   Check live copy against the dead copy.

Discussion:

Proofread after you have checked for correctness, when you think the document is standard and consistent, to find overlooked errors or errors that crept into the current (live) copy.

Proofreading, the final step in the writing system, demands patience and disciplined attention to detail.

Often, poor proofreading occurs because writers try to combine editing with proofreading. Consequently, they lose focus.

Technique 13.1   Proofread in a series of readings.

Tips:	In a series of readings, consider word choice, grammar, punctuation, mechanics, typing, and content.
	Use proofreading tools: spellchecking software, dictionary, style guide, and a straight-edge to check columns and alignment.
	Mark corrections in the text, then note them in the margin. Count the corrections per page and put the number at the top right corner of the page.
Warning:	Do not try to check everything at once.
	Do not revisit editing steps as part of your proofreading. If you do, you introduce errors.
Example:	Look for these common typing errors:

- Incorrect word break (Most software packages use algorithms to break and hyphenate words. The algorithms make mistakes.)
- Duplicated or omitted letters, words, or lines
- Mistyped numbers in dates, statistics, formulas, or money
- Missing punctuation marks that usually come in pairs (commas, parentheses, quotation marks, dashes)
- Typos in material outside the body text (titles, captions, table of contents, headings, addresses, signature lines)
- Faulty spellchecker word substitution
- Transposed letters—*form* instead of *from*

Look for these common content errors:

- Inaccurate page citation—sending reader to wrong page
- Inaccurate table of contents and index
- Inaccurate detail (Baltimore, Virginia's chief port city)

See also:	Correctness.

Discussion:

Print out your electronic document for proofreading. Paper documents are easier to check; therefore, accuracy improves.

Proofread slowly—more slowly when the content is technical, in outline format, or uses varied typefaces and point sizes. Be careful at the end of long lines and in the middle of long words. If you find one error, look back for others—errors often cluster.

Be aware of your own weaknesses. Do you commonly transpose letters? repeat words? Take breaks to stay alert.

Exercise: Proofread the following letter. Identify errors as word choice, grammar, punctuation, mechanics, typing, or content. (Answer A-56)

12 June 1991

Graphics Leasing Corp.
Attn: Accts. Receivable Manager
VGS Park Dept A
5701 N.W. 9th Ave.

Dear Sir:

Below is a list of Alcor Corp. check numbers, invoice dates and amounts which are in payment of the lease and maintenance agreement on the POS one 320 camera presently in our possession. This information was obtained through our disbursement summery report for fiscal year 1981.

Becuase of age of these items, neither cancelled checks nor duplicate copies can be obtained for review. I suggest, that you check your deposit records for the time in question to varify the amounts.

Check #	Invoice Date	Amount paid
037614	4/25/89	$ 932.04
043569	4/25/89	804.12
0128476	4/28/89	804.04
019383	5/14/89	425.12
019383	5/14/89	425.12
036000	7/3/89	445.12
036001	8/15/98	388.24
037455	0/11/89	432.17
045672	11/22/89	464.25

Total Paid $ 5,110.25

These problem have occured because of your inaccurate files. The discrepency is seven and one half and one half years old, therefore, we are not responsibile from proving any fa-rther the payment of these items. I consider this matter closed with this letter and the schedule of paid items.

Sincerely

Accounting Supervisor

Technique 13.2   Check live copy against the dead copy.

Tips:
Make sure your live copy includes corrections marked on the dead copy. Count corrections and compare to the total recorded at the top right corner of the dead copy.

Make sure no errors crept in. Mark or disfigure the dead copy and put it away when you are through checking so dead copy doesn't find its way back into the live document.

Check legibility, the quality of type—faded? smeared? dark? streaking? wrinkled paper? misaligned paper or any other printer mechanical problems? Check to make sure you have adequate margins for binding, and clear, crisp graphics.

Check completeness: page sequence, auxiliary material (such as appendices), enclosures (such as photos).

Warning:
Do not proofread if tired or in a hurry. Hurried proofreading tends to introduce errors. Plan time for proofreading.

Do not assume that boilerplate has been proofread and is therefore correct.

Example:

Dead copy: Pete, Don and Mike met today to discuss out Spring schedule.
Live copy: Pete, Don,and Mike met today to discuss our spring schedule.
Note: Proofreader found and corrected three errors in the dead copy (an omitted comma, misspelled word, and unnecessary capital), but introduced a new error to the live copy (an omitted space).

See also:
Correctness.

Discussion:
Editors refer to the current copy as "live copy" to distinguish it from the previous "dead" copy. Establish procedures to keep track of your proofreading. Otherwise, you waste time through unnecessary re-reading, or you forget to make the corrections marked.

Proofreading ensures the physical quality of the document. Check the paper, type, layout, graphics, and binding. Even the best laser printers occasionally double-feed sheets, wrinkle paper, put pages out of order, streak, and fade. Ensure everything is included in the right order, with nothing duplicated.

Exercise: Check live copy against the dead copy. What errors were corrected? What errors remain? Did new errors creep in? (Answer A-57)

Dead copy	**Live copy**

Dead copy

Before beginning this tutorial for the System Administrator of the Blue System, be sure that your have already finished the tutorial for the Tech Controller. That tutorial lays the foundation on which all Blue System tutorials are based. This tutorial describes only those administrative tasks that are relevant to the Blue System operated on a SUN Workstation.

Also, before beginning this tutorial, be sure that you are comfortable using the *vi* editor described in detail in Chapter 6 of *Getting Stated with UNIX—Beginner's Guide,* available from your instructor. Recall that some men have items that are not selectable, because they are limited to user's with certain privileges, who are
    1. System Administrator
    2. Circuit Connectivity Modifier
    3. Pin Connectivity Modifier
    4. Technical Controller.

A privileged user may access only certain windows. An individual who has been assigned only System Administration privileges, will be able to open only the System Administration Window, the Console Window, the Message Window, and the RED Messages Window. This tutorial for System Administrators assumes that you have been assigned only System Administrator privileges. This tutorial will describe each window that you you may access, its functions and tasks.

Live copy

Before beginning this tutorial for the System Administrator of the Blue System, be sure that your have already finished the tutorial for the Tech Controller. That tutorial lays the foundation on which all Blue System tutorials are based. This tutorial describes only those administrative tasks that are relevant to the Blue System operated on a SUN Workstation.

Also, before beginning this tutorial, be sure that you are comfortable using the *vi* editor described in detail in Chapter 6 of *Getting Started with UNIX—Beginner's Guide,* available from your instructor. Recall that some menues have items that are not selectable, because they are limited to users with certain privileges, who are
    1. System Administrator
    2. Circuit Connectivity Modifier
    3. Pin Connectivity Modifier
    4. Technical Controller

A privileged user may access only certain windows. An individual who has been assigned only System Administration privilegeswill be able to open only the System Administration Window, the Console Window, the Message Window, and the RED Messages Window. This tutorial for System Administrators assumes that you have been assigned only System Administrator privileges. This tutorial will describe each window that you you may access, its functions and tasks.

Step 13. Proofread

Final exercise: Proofread this memorandum. (Answer A-57)

INTEROFFICE CORESPONDANCE

To: Distributoin
From: Mary Poole
Date: 2 Setpember, 1991

Re: Briefing for Training Faciltators

You have been assigned training resp onsibilities in conjunction with this years
MGR training effort. The two short tapes which will be used for the the 1991
training willbe "Timekeeping" and "Harrassment. The tapes were all ready in
production and will be   shipped at you approximately October 1.

A briefing session will be held 11 a.m on June 28, 1990 at headqurters to reveiw
the to review the content of the tapes, identify expected discussion topics, and to
provide some direction for leading the discussion periods. I look forward to seeing at the meeting. Although we may extent into the noon hour, you may exspect
that we will be finished by 2 P.M..

You should move forward with scheduling trianing sessions at your employees locations,
   with goal of completing all training bythe end of Auguest.

Distribution:
      Presedent
      Controllers Office
        All Senior technecal staff

Final Final exercise: Congratulations! You've worked through the Writing
System. In conclusion, describe how mastery of the Writing System helps you on
the job. (Answer A-58)

Appendix A

Answers to Exercises

Step 1. Analyze Purpose

(From p.7) Exercise: List and contrast three purposes associated with each writing task.

1. You must write a required monthly status report to your client, the City of Boston, who hired your company to clean the water in Boston Harbor.
   - Your purpose for writing: inform about accomplishments, pending work, issues
   - Reader's purpose for reading: monitor work and provide feedback
   - Purpose of the work: clean Boston Harbor water

2. After receiving many phone queries, you decide to write a memo to your plant employees describing procedures for getting their monthly parking sticker from their shift supervisor.
   - Your purpose for writing: provide accurate guidance so you won't be bothered with calls
   - Reader's purpose for reading: learn how to get a parking sticker
   - Purpose of the work: control parking at the plant

3. As office manager, you write a staff study to your immediate supervisor recommending leasing rather than buying a copy machine.
   - Your purpose for writing: explain why leasing the copy machine is better than buying
   - Reader's purpose for reading: evaluate your recommendation, then approve or disapprove
   - Purpose of the work: provide cost-effective copying for the office

(From p. 8) Final exercise: List and contrast three purposes associated with this writing task.

   - Your purpose for writing: boost sales and get rewards.
   - Store manager's purpose for reading: carry out marketing instructions.
   - Purpose of the work: stock shelves and take advantage of promotional campaigns.

Step 2. Analyze Audience

(From p. 11) Exercise: Fill in the blank to identify the audience matching each set of characteristics.

1. **Operator** This audience needs precision. They need to know what and when more than why. They want to know—What procedures must I follow to begin, continue, and complete a job? What equipment must I install or operate? What does it look like and how does it work? Are alternate procedures useful or possible? What can go wrong, and how do I correct if it does?

2. **Layperson** This audience wants the interesting or useful nuggets of information. They read for curiosity or self-interest. They won't read technical language and theories. They consider many details unnecessary. They want you to translate jargon into standard English, using short words and short sentences when you discuss difficult subjects. They need analogies and examples to understand your message. They need just enough background to follow your discussion easily.

3. **Manager** This audience wants a concise presentation of practical information. They want to know—Is this proposal feasible? What will it cost? Is it cost effective? Who is doing it? How well is it being done? Are things running on schedule? What's going to happen next? What advantages will result for the organization? What do you want me to do?

4. **Expert** This audience requires detail. They may want to know why as well as what. They want to know—What problems need to be solved? What procedures will be used? Does the product meet design and performance specifications? Is there adequate and documented data to support conclusions? How can these conclusions help us in further research, product development, or training?

(From p.13) Exercise: List and profile audiences.

Audience	**Store Mgrs.**	**Floor clerks**
Priority	*Primary*	*Secondary*
Reading as	operator	operator
Profession	management	sales
Knowledge of subject	technical	general
Friendly, hostile, neutral	friendly	neutral
Considers you expert	yes	yes
Wants background	no	no
Wants theory	no	no
Wants examples, pictures	no	no

(From p.14) Final exercise: This memo's writer skipped analyzing purpose and audience.

1. Analyze purpose:
   - Writer's purpose for writing: report findings on TaskPert Software.
   - Reader's purpose for reading: find out if the company is buying TaskPert Software.
   - Purpose of the work: evaluate TaskPert Software.

2. Analyze audience:

Audience	**V.P. Strategic Planning Brown**
Priority	*Primary and only*
Reading as	manager
Profession	business
Knowledge of subject	functional
Friendly, hostile, neutral	friendly to neutral
Considers you expert	yes
Wants background	no
Wants theory	no
Wants examples, pictures	no

3. Recommend changes to the memo:

Writer's purpose dominates the memo and is at odds with the reader's purpose. Writer leads the reader through details that the reader doesn't want, then reaches a conclusion at the end of the memo. Put the conclusion up front. Either cut the details or put the details in an attachment, perhaps a fact sheet. Then, the writer has a record of findings that does not intrude on the reader's simple interest in the conclusion: are they buying the software?

Step 3. Write a Purpose Statement

(From p. 18) Exercise: Write a purpose statement after analyzing purpose and audience.

Scenario 1:

1. Analyze purpose:

- Writer's purpose for writing: avoid the problems of towing, fees, and angry employees.
- Reader's purpose for reading: learn where and when to park and not park.
- Purpose of the work: maintain the parking lot.

2. Analyze audience:

Your audience is all employees who use the parking lot. They read as operators: that is, they just want to know what to do and when to do it. They have little interest in background or theory or parking lot maintenance, but they might profit from a diagram.

3. Purpose statement:

memo	notifies	employees	restrict parking	plan accordingly
Actor	Action	Audience	Object	Outcome

This memo notifies employees who use the parking lot of restricted parking next week so you can make other plans.

Note: We chose *notifies* to set a formal and somewhat authoritative tone. You may prefer a less authoritative tone such as *inform*.

Scenario 2:

1. Analyze purpose:

- Your purpose for writing: get the $98,000 additional work.
- Reader's purpose for reading: decide to buy the bulletin board function.
- Purpose of the work: expand the E-mail system to include a bulletin board.

2. Analyze audience:

Your audience is the client who reads as a manager: that is, the client must decide. The client wants cost benefit analysis, maybe some background, but no theory about E-mail.

3. Purpose statement:

change proposal	outlines	the client	benefits-cost	decide to add
Actor	Action	Audience	Object	Outcome

This change proposal outlines to the client the features, benefits, costs, and schedule for adding a bulletin board function so the client can decide whether to add the function.

(From p.19) Exercise : Evaluate these purpose statements. Are all five parts included? Do the five parts clearly and logically state the document's purpose?

1. This magazine article (*actor*) explains (*action*) to the general reader (*audience*) the advantages of recycling paper (*object*) so residents know how to cooperate in the county's recycling program (*outcome*).

Clear but not logical. The outcome *know how to cooperate* doesn't result from explaining advantages.

2. This user guide informs new employees about procedures for using E-mail so they can send and receive E-mail.

    Actor—user guide
    Action—informs
    Audience—new employees
    Object—procedures for using E-mail
    Outcome—so they can send and receive
    Clear and logical.

3. This proposal explains our technical approach for the client users, and it presents our concept for the management plan, schedule, and costs for the financial department, and it justifies our bid in terms of time and materials, so you have the necessary information to evaluate our qualifications to successfully complete your project.

Actor—proposal		
Action—explains	presents	justifies
Audience—client users	finance dept.	you (understood)
Object—approach	plan, schedule, costs	bid
Outcome—so you can evaluate our qualifications		

Not clear, not logical. We have four purpose statements merged into one, and each of the four is missing parts of a purpose statement. 1. We have a section explaining to users our technical approach with no apparent outcome. 2. We have a section presenting to the finance department the plan, schedule, and cost with no apparent outcome. 3. We have a section justifying to reader (you understood) 4. We probably have a section that helps the reader evaluate our qualifications, perhaps a discussion of our experience and resumes.

4. I thought I'd like to write down some thoughts and concerns for you in regard to the TDMS as we approach trials. We need to be constantly aware of several things, monitor their progress, and/or verify the fix.

    Actor—omitted
    Action—omitted
    Audience—you (understood)
    Object—thoughts and concerns—*vague*
    Outcome—omitted, or possibly so you can be aware, monitor, and fix?
    Not clear. Fill in omissions.

5. This letter is a follow-up to our phone conversation today regarding the above captioned.

    Actor—letter
    Action—follows up
    Audience—you (understood)
    Object—omitted
    Outcome—omitted
    Not clear. Fill in omissions.

Answers to Exercises

(From p.21) Exercise: Help B. Guirre devise a purpose statement for this memo.

Purpose statement:

This memo reports to ACME Temps your employee's poor performance so you can investigate her behavior.

> or—so you can discipline her?
> or—so you can make amends?

How does a purpose statement focus B. Guirre?

B. Guirre doesn't seem to know what she wants. In the letter B. Guirre just lets off steam. Here's what I tell B. Guirre: Why write a letter without a clear purpose? A purpose statement forces you to decide if you have a point worth making. Do you want a refund? Do you want an apology? Do you want revenge? Maybe after analyzing your purpose you'll decide not to write such letters. Ida already wasted your weekend. Now the ghost of Ida is going to waste your workday morning.

(From p.22) Final exercise: Analyze purpose and audience, then write a purpose statement for the system design.

1. Analyze purpose:

- Your purpose for writing: provide sufficient detail to build the fulfillment and invoicing system for Pinnacle Trail Ltd.
- Your technical staff's purpose for reading: write code to modify the UNIX-based package. IS Manager's purpose for reading: approve the system design. VP of Sales's purpose for reading: see how new system affects his operation.
- Purpose of the work: automate fulfillment and invoicing.

2. Analyze audience:

Note: Whomever you select as the primary audience determines your purpose statement. We chose our technical staff as the primary audience. You may disagree.

Audience	**Tech staff**	**IS Mgr.**	**VP Sales**
Priority	*Primary*	*Secondary*	*tertiary*
Reading as	operator	manager	layperson
Profession	software	info. mgt.	sales
Knowledge of subject	technical	technical	minimal
Friendly, hostile, neutral	friendly	neutral	hostile
Considers you expert	yes	yes	yes
Wants background	no	some	yes
Wants theory	no	some	no
Wants examples, pictures	yes	yes	yes

Purpose statement: This system design specifies to the technical staff the modifications required to the UNIX-based software package so the staff can write, compile, and test code.

Note: You must write to your primary audience. You can see that you cannot simultaneously satisfy all three audiences. Later, in Edit for Coherence, you learn techniques to make your document accessible to your secondary and tertiary audiences.

Step 4. Gather Information

(From p.25) Exercise: Describe how the following purpose statements focus information gathering.

1. This bid request specifies to ACME Carpet Co. the dimensions of eight vacant offices at 123 Maple Avenue so your carpet company can bid on installing wall-to-wall carpeting.

    Actor—bid request implies first step toward contract
    Action—suggests precise details
    Audience—ACME Carpet, reading as manager, knowledgeable and friendly audience
    Object— suggests eight groups of measurements
    Outcome—suggests a formal reply describing work and price

2. This brochure outlines for prospective renters the features and benefits of our vacant offices at 123 Maple Avenue so prospective renters can decide whether to schedule an appointment to view the office space.

    Actor—suggests a marketing piece
    Action—suggests a general discussion, informal, less precision
    Audience—suggest people who already know they want rental space
    Object— suggests a limited discussion of the offices' good points, not costs
    Outcome—suggests that brochure need only say enough to entice prospect to call

(From p.27) Exercise: Ask *who, what, where, when, why,* and *how* questions to generate details for documents with these purpose statements.

For each document, we thought of questions—you may have different questions.

1. This tech-report describes the repairs and tests we conducted on your PDQ Laserprinter to stop the intermittent errors you reported, so you know what was covered by warranty and what service you must pay for.

Who repaired and tested? Whom do you call for more help? What repairs? What errors? What warranty? What do we check next if intermittent errors continue? Where did we make the repairs—shop or on site? When did we receive the call? make the repairs? When will you get the invoice? When is payment due? Why are some repairs under warranty, others not? How will we know if the repairs are effective?

2. This guide provides you, the new owner of a Megawheel plastic tricycle, simple step-by-step assembly instructions so you can quickly put your little tyke on his Megawheel trike.

Who manufactured trike? Who assembles trike? Whom do I call for help? What tools do I need? What time do I need—really? Where is the bag of nuts and bolts? When do your tech-support staff answer calls? Why are there eight bolts and seven wing nuts? How do I start? How do I know when I'm finished? (Lots of questions need to be answered here.)

3. This letter notifies LD Cellular Phone Co. billing office about $12,456.56 of unauthorized calls charged to our account so you can change our phone access code, investigate the fraudulent calls, and remove the charges from our account.

Who is responsible for unauthorized calls? What measures did we take to avoid this form of theft? Where were the calls made to? When did we notice the unauthorized calls? Why are we asking you to change our phone access code? How do we prove the calls are unauthorized?

Answers to Exercises

(From p.28) Final exercise: Use your purpose statement and the questions *who, what, where, when, why,* and *how* (the 5 Ws and H) to focus information gathering.

Scenario: You are a regional sales manager for a rapidly expanding chain of hardware stores. You send a monthly report. Your purpose statement reads: This monthly marketing report informs you of new products, prices, and promotional efforts so you can prepare your store to take maximum advantage of sales opportunities.

1. How does the purpose statement focus your gathering information? Also, give an example of data you don't need to gather.
> Actor—monthly report—suggests a regular format requiring timely action
> Action—suggests formal communication, not quite a command
> Audience—store manager, reading as operator, knowledgeable and cooperative audience
> Object— suggests precise information about related topics: product, price, and promotion
> Outcome—suggests benefit or reward for acting on information in report

You don't need to gather persuasive details, background, or known policies and procedures.

2. Using the 5 Ws and H, what information do you gather?

Who is running the sales promotions? Who will send the promotion kits and merchandise? What products are featured this month? What local promotion must I do? Where do I put the special displays? Where do I send unsold goods? When does the promotion start and stop? Why are we running this promotion this month? How do I order more merchandise if I run out?

Step 5. Write Sentence Outline

(From p.31) Exercise: Using gathered information as your source, write assertions for a sentence outline. Use short words in short sentences.

Information gathered:

1. *Who* applied for job sharing? Twenty BAP employees and ten non-employees applied for the job sharing pilot program. Applicants varied from a part-time attorney to a full-time senior manager. Most applicants were administration and human resource department professionals with a range of experience.
Assertion: Many competent applicants with a wide range of experience applied.

2. *What* are the advantages and disadvantages of job sharing? The corporation retains good employees who know the job and the organization. Job sharing entails little extra cost. Job sharing increases employee morale. Management can lose control. Customers can become confused. Job sharers can be incompatible.
Assertion: Job sharing advantages outweigh the disadvantages,
Or: Job sharing has three key advantages and three key disadvantages.

3. *Where* does job sharing occur? The percentage of U.S. companies that job share by region: West–8%; Mid-west–13%; South–6%; Northeast–7%; Great Lakes–16%.
Assertion: Companies across the United States use job sharing.
Or: Companies in the Great Lakes region or Mid-west are twice as likely to job share as the rest of the U.S.

4. *When* do job sharers work? Each partner works 20 hours a week and earns 1/2 company benefits. Each partner works a $2\frac{1}{2}$ day schedule. Partners work 2 hours together mid-week to ensure continuity.
Assertion: Job sharers set a schedule to satisfy their needs and the company's needs.
Or: Each job sharing partner works a half-week.

5. *Why* did applicants want to job share? Some wanted time to pursue personal interests like education. Some working parents wanted more family time.
Assertion: Applicants offered different reasons for wanting to job share.

6. *How* did job sharers communicate the new working arrangement to others? They published internal memos introducing the change. They sent a letter to vendors and clients. They reminded contacts of the job sharing arrangement when using voice mail, telephone, or written communication.
Assertion: Job sharers carefully communicated their new working arrangement to others.
Or: Job sharers used every opportunity to communicate their new working arrangement to others.

Answers to Exercises

(From p.33) Exercise: Evaluate the following assertions against the purpose statement. Eliminate irrelevancies and redundancies. Tip: Concentrate on costs and benefits so the audience can decide whether to invest.

Purpose statement: This fact sheet highlights for senior managers the costs and benefits of our proposed automated data security system so you can decide whether to invest in the new system.

Assertions:

~~Ninety percent of Fortune 500 companies use data security systems.~~ (irrelevant)

~~Our professionals keep a lot of valuable information on our computers~~. (redundant)

We rely on computers now more than ever to be profitable.

~~Losing data can severely reduce profits.~~ (redundant)

The proposed system uses three optical drives tied to our local area network.

~~The next-best alternative used old technology, a 16 PI tape drive~~. (irrelevant)

The 3 optical drives, 10 cartridges, optic fiber cables, and software cost $6,490.

~~Causes for data loss range from employee error to natural disaster.~~ (irrelevant)

~~Industry surveys provide statistics on industry-wide information loss~~. (redundant)

The proposed system limits data loss to a worst case 24 hours.

Last year, lost data cost our company over 3,500 man hours at an average $40 per man hour.

Our costs for losing data exceeded industry averages.

Expect our proposed data security system to reduce data losses by 85%.

Lost data results in misplaced or late orders, hence angry customers.

(From p.35) Exercise: Determine the main assertions and supporting assertions. Group supporting assertions under main assertions. Put groups in order.

Scenario: Your New Jersey company plans to build a can manufacturing plant somewhere in the Southeast. You must write a staff study to the company president showing the pros and cons of building the plant in Tuscaloosa, AL.

Purpose statement: This staff study details to you (the president) the pros and cons of building a can manufacturing plant in Tuscaloosa, AL, to help you decide where to locate the new plant.

2. At present, Tuscaloosa's business climate meets our needs.

    9. Recent layoffs in local chemical, rubber, paper, and iron factories make an abundance of cheap, skilled labor.

    7. The local area costs are low for our business.

    12. State and local governments offer five-year tax incentives to move into Alabama to prop up the stagnant economy.

14. Tuscaloosa's transportation is geared for manufacturing companies.

    15. Tuscaloosa serves as a minor hub for both highway and rail traffic.

    1. The Black Warrior River-Tombigbee Waterway links Tuscaloosa to the Port of Mobile.

    6. The Tuscaloosa Airport requires connections through Atlanta.

16. Tuscaloosa presents a major change in living environment for our New Jersey transplants.

    3. Tuscaloosa provides few big-city amenities.

    13. Local public schools are rated below average, but improving.

    8. Personal income, property, and sales taxes are low.

    4. Tuscaloosa has no big-city costs.

    5. The University of Alabama offers some cultural opportunities.

    11. Tuscaloosa has mild winters and hot summers.

    10. Tuscaloosa has many recreational facilities: lakes and parks.

Answers to Exercises

(From p.37) Exercise: Identify the natural patterns used to order each group of assertions and details.

Group 1   Topical

Group 2   Functional

Group 3   Chronological

(From p.38) Final exercise 1: Write a sentence outline.

1. Analyze purpose:
   - Your purpose for writing: dissuade your client from accelerating implementation, or at least warn of the risks and insulate yourself from a potential disaster.
   - IS manager's purpose for reading: learn your assessment of the risks.
   - Purpose of the work: upgrade the transaction and billing system.

2. Analyze audience: Your primary audience is the IS manager who must evaluate your assessment of the risks and decide whether to push for accelerated development. The IS manager wants the accelerated development and may be hostile to your message. However, the IS manager considers you an expert.

3. Purpose statement: This letter describes the technical and business risks associated with accelerated, November, deployment of your Bank Card system so you can factor the risks into your decision on accelerating deployment.

4. Sentence Outline:

1. We recommend you stay with the original schedule to avoid technical and business risks.

2. The technical risks happen because we cannot accelerate delivery unless we shorten testing.

3. You face two business risks if you deploy the system during the Christmas retail season.

      3.1 Unreliable performance during 45-day shakedown might cost $1.3 million in lost revenue.

      3.2 Unreliable performance during 45-day shakedown will hurt your customers.

4. We believe the potential $120,000 savings in November do not justify these risks.

(From p.39) Final exercise 2: Use your sentence outlining skills to disassemble and then reassemble this poorly organized letter.

First, write each assertion you find in the letter as a short sentence.

1. Thank you for your good service.

2. Some suppliers give customers gifts during the holidays.

3. We think suppliers should not provide gifts.

4. We don't think you're crooked.

5. We want you to spend your money improving your services.

6. Thank you for not giving us gifts.

7. Happy Holidays.

Second, analyze purpose and audience.

- Writer's purpose for writing: encourage compliance with company policy.
- Reader's purpose for reading: find out if they can send gifts.
- Purpose of the work: prevent favoritism.

Audience is suppliers who want to maintain good relations with the company. They may or may not know company policy concerning gifts. They are open to the message, and do not need or want background information to follow instructions.

Third, write a purpose statement for the letter. (Don't try to make your purpose statement accommodate all the original assertions.)

This letter informs our suppliers that we cannot accept holiday gifts for your information.

Fourth, using your purpose statement, evaluate and order assertions in a sentence outline.

1. Please do not send gifts (omitted in original letter).

2. Thank you for your good service.

3. Happy Holidays.

Other assertions are irrelevant (even insulting) or redundant.

Step 7. Revise Content and Organization

(From p.47) Exercise: Apply the Content Test to this letter. Suggest ways to improve content. Do key words in the first two sentences make the *topic* clear? Delete assertions that fail the *so what?* test. Suggest supporting details where an assertion needs to *specify how*.

Question 1. What is the topic?	We added key words: disputed transaction on VISA bill.
Question 2. So what?	We crossed out assertions that failed *so what?* test.
Question 3. How supported?	We italicized the unsupported statement.

This letter informs you of a disputed transaction on our VISA bill, so you can investigate and adjust the billing error. The transaction that I have circled is the transaction we are disputing. The date of the transaction was August 5th and the date it was posted was August 8th. The company was Greenspan & Co., Inc. in Tuscaloosa, Alabama. ~~I have never heard of this company before.~~ The amount charged was $520.97. ~~That is why I am questioning this transaction. Is there any more information that you can send us regarding this transaction?~~

~~Our accounts payable department and I have no idea where this bill has come from. We have a purchase order system and I would have known if I had approved any request involving the Greenspan company. I didn't. It is possible that our number was mistakenly used or punched into the computer. I personally know how easy it is to transpose numbers.~~ *Your help in any way would be very appreciated in this matter.* Change to *Remove the $520.97 charge from my account before the next billing cycle.*

Sincerely,

(From p.49) Exercise: Apply the Organization Test to this letter. What advice can you give?

Question 1. Reads like a data dump?	No.
Question 2. Reads like a story?	Yes. *I learned about job, got excited. . . .*
Question 3. Filled with *I, me, mine*?	Yes. Letter tells how writer is affected.

Joan can't edit her way out of this problem. She needs to analyze purpose and audience, write a purpose statement, then write a brief sentence outline. She focuses on her purpose, but ignores the reader's purpose. Consequently, the reader will likely ignore or be put off by her letter.

Remember, when applying for a job, the purpose of your cover letter is to get your reader to look at your resume; the purpose of your resume is to get an interview; and the purpose of your interview is to get an offer.

(From p.50) Final exercise: Apply the Content and Organization Tests. Recommend ways to improve the letter.

Content Test:

Question 1. What is the topic?	Not stated at the beginning of the letter.
Question 2. So what?	We crossed out assertions that failed *so what?* test.
Question 3. How supported?	We italicized the unsupported statement.

Organization Test:

Question 1. Reads like a data dump?	No.
Question 2. Reads like a story?	Yes. The life and hard times of my battery. . . .
Question 3. Filled with *I, me, mine*?	Yes. Letter tells how writer is affected.

On December 21 of last year, we asked for reimbursement for an alternator repair ~~to our new company car, which we purchased through a dealer who sold us~~ your extended warranty. (See attached letter.) *Now as a result of the defective alternator, the company car's battery died, and had to be replaced. (Specify which alternator: the former or the latter?)*

When the service center replaced my alternator, they had trouble recharging the battery, but they suggested I keep the old battery for awhile to see if it would hold a charge over a period of time. As it turned out, two weeks after the alternator was replaced, the battery died. ~~Although it's December, the weather has been unseasonably mild; so I guess cold weather did not kill the battery~~. The battery died as a direct result of the previous defective alternator running it down over a period of time.

~~I realize that the battery is not specifically covered under our warranty, but~~ because it died as a direct result of the mal-functioning alternator, ~~I think it's only fair that~~ you replace the battery. ~~It was only 17 months old. We shouldn't have had alternator or battery problems in the first place.~~ *Please call me if you have any problems with this. (Specify how and when.)*

Recommend ways to improve the letter:

Write a purpose statement. Then disassemble the main points of the letter and reassemble using sentence outlining techniques. Always close business letters with a specific call to action.

Step 8. Edit for Coherence

(From p. 53) Exercise: Shifting words make this memo written by a software engineer difficult to follow. Improve coherence by identifying shifting words and replacing them with consistent key words.

Shifting words: For each set of shifting words, we selected a consistent key word. You may select other key words.

concerns, thoughts, issues, questions = concerns

end user, operator, user, customer = end user

capability, ability, feature = capability

commercial graphics package, standard software packages, graphics packages = commercial graphics software packages

Subject: ReportMaker Database *Concerns*

These Reportmaker Database *concerns* are for your information and comments. I tried to define some *concerns* we've been grappling with as we focus on the *end user's* needs. I met with Kathy Barlowe on May 10 to discuss some of the *concerns* involved in supporting *commercial graphics software packages* with the database. Kathy and I tried to better define *end users* who may call for *capability* to incorporate graphics into ReportMaker.

Software Compatibility *Concerns*

If we plan to offer this *capability* to the *end user*, we need to test this *capability* for *commercial graphics software packages* and list guidelines for the *end user*. For example, if only a few *commercial graphics software packages* can be pulled into ReportMaker, we should say so to take out the guesswork.

Note: The key word repetition emphasizes the key points of the memo: concerns, end users, capabilities, and commercial graphic software packages.

(From p. 53) Exercise: Improve this excerpt from a technical report by identifying shifting words and replacing them with consistent key words.

Shifting words: Don't worry if your choice of key words differs from ours. Concentrate on identifying and grouping the shifting words. Then choose a key word to replace the shifting words.

Office of Public Works, Public Works = Office of Public Works

employees, workers, peers = employees

solution, methods, approaches, course of action = solution

pro's and con's, advantages and disadvantages = advantages and disadvantages

dedicated modem, fax modem = fax modem

fax servers, PC server, server = fax servers (Change *dedicated PC server* to *dedicated PC*.)

fax traffic, fax volume = fax volume (Let *fax volume* contrast with *network traffic*.)

The *Office of Public Works* needs to implement a fax *solution* that enables *employees* to send and receive faxes from and to their computer workstations. The *Office of Public Works* also wants *employees* to be able to share faxes electronically.

Fax *Solutions*
We can use one of the two broad *solutions* of implementing PC-based fax capabilities in mid-sized offices given our moderate *fax volume*. The table below illustrates the two *solutions* and their *advantages* and *disadvantages*:

Solution	*Advantages*	*Disadvantages*
Fax Modems	Handle large *fax volume*	Very expensive
	Provides flexibility	No central control
Fax Servers	Provides central control	Requires *dedicated PC*
	Shares resources	Increases *network traffic*

Because of the expense associated with providing a *fax modem* per user, we do not recommend that *solution*, but prefer the *fax server solution*.

Note: The key word repetition emphasizes the key points of the memo: Office of Public Works, employees, solution, advantages and disadvantages, fax modem, fax server, fax traffic, and fax volume.

Answers to Exercises

(From p.55) Exercise: The key sentence tells the reader the purpose of the paragraph. A coherent paragraph puts the key sentence first for emphasis, then develops it logically with following sentences. Why are these paragraphs incoherent?

Paragraph 1: The key sentence says we're *comparing similarities*. Then the developing sentences *contrast differences.*

The TAKKS-C architecture is **similar** to TAKKIMS in that the local area networks are comprised of fiber rings, although the TAKKS-C fiber optic receivers are single fiber devices and less fault-tolerant than the dual port transceivers proposed for TAKKIMS. Another **distinct difference** between TAKKS-C and TAKKIMS is the interconnection of the sites. TAKKS-C used ports on the host computer to interconnect geographically separated LANs. In TAKKIMS, gateways attached directly to the fiber optic LANs will provide a network which appears to the users and software as a single logical network. **Unlike** TAKKS-C, TAKKIMS will provide fiber optic LANs at all remote sites.

Paragraph 2: The key sentence indicates we discuss first closing the account, then secondly referring the account. However, the developing sentences reverse the order. Also, the paragraph shifts key terms: retention, saving.

CATS Representatives respond to telephone close requests by either **closing the Customer's account** or **referring the Customer to the Customer Retention department**. The Flag field on MBNAIS indicates to the Representative if the Customer is profitable and worth saving. If the account is profitable and the call is taken during Customer Retention's operating hours, the CATS Representative will transfer the Customer to Retention. If the Customer Retention is closed, the CATS Representative will attempt to save the account by explaining the benefits of the MBNA card and possibly offering a different interest rate (associate bank charge.) If the account is not profitable, the CATS Representative will close the account and make appropriate monetary adjustments.

A-18

(From p.57) Exercise: Use vertical lists to logically group information for the reader.

1. Follow these steps to import line graphs when word processing:

    a. Open an empty frame where you want to put your line graph.

    b. Choose Graphics, Figure, Retrieve from the menu bar.

    c. Select and import line graphs from the menu by double-clicking on the desired filename. (Your selected line graph appears in the frame.)

    d. Adjust the position or size of the frame by clicking and dragging a frame handle.

    e. Anchor your frame to the page by double-clicking the anchor icon.

2. Disbursement requests must be properly documented for payment processing. Each disbursement request must contain

- payment authorization
- invoices signed by the approving department
- receiving copy of the purchase order
- other receipts such as shipping documents
- remittance for any merchandise returned

The disbursement system allows some unique purchases on a case-by-case basis. For example, hardware maintenance outside our standard maintenance contracts requires a copy of the estimate and final bill. Advertising requires proof-of-service such as a copy of the advertisement, the order, and authorization.

(From p.59) Exercise: Break this long paragraph into shorter paragraphs.

ARNEWS is the bi-monthly newsletter of the Chief, Army Reserve. The Public Affairs office publishes the contents, current events in the Army Reserve. In addition, Public Affairs maintains the mailing list for ARNEWS's subscribers.

To create an edition of ARNEWS, the Public Affairs office collects the articles that comprise the newsletter. Articles come from sources within the Army Reserve as well as outside sources contributing for publication. Public Affairs edits the articles and desk-top publishes a camera-ready copy of the newsletter.

To create a mailing list for ARNEWS distribution, Public Affairs downloads from its membership database an ASCII file containing name, address, zipcode, and member ID. The ASCII file has one record for each line, followed by a carriage return. The record fields are delimited by commas.

Public Affairs sends the camera-ready copy of the newsletter and a 16 BPI tape of the mailing list to the publisher. In turn, the publisher prints the newsletter, creates and affixes cheshire labels, then mails the newsletter.

Answers to Exercises

(From p.61) Exercise: This introduction lacks one of the seven parts. Identify the six parts it has and the one missing.

1. Purpose statement: missing

2. Organization of document: . . .*optimum flowchart layout of the monthly cycle. Each step has a corresponding chapter. Follow the cycle each month and refer to the appropriate chapter as you work through this manual.*

3. Background and significance of topic: *Each month company executives view a set of charts, or slides, which have more current information than those of the previous month. Very basically, the new data comes in and is checked, then slides are generated. . . . .*

4. Description of target audience: *designed specifically for Executive Support Personnel in the Product Support area of BAP, Inc. . . .*

5. Information sources and research methods: *Oracle software, Generic Graphics, manuals in the Product Services library. . . .*

6. Definitions of key terms: *The set of characters <RETURN> stands for Carriage Return and is simply a tap on the key labeled "Return" on your keyboard . FDBA stands for Functional Database Administrator.*

7. Limitations of the document: *This manual assumes that the reader has a working knowledge of the Slide Presentation system in the VAX environment. . . .*

(From p.63) Exercise: Which of the following two passages is an abstract, and which is an executive summary? How can you tell?

Mars and Luna Direct is an abstract that highlights key words from the paper and uses technical jargon to summarize purpose, methods, and findings.

Losing the High Frontier is an executive summary that provides some background, states the main assertion of the article, then proposes a practical consortium to devise a blueprint for space. Language is non-technical.

(From p.65) Exercise: Remember, a document with zero coherence devices is just one huge block paragraph. The following passage is well organized but needs visual coherence devices to make it reader-friendly. Add visual devices.

TO:	Senior Management
FROM:	John Phelps
DATE:	May 3, 1991
SUBJECT:	Long Distance Telephone Credit Cards

Effective immediately, BAP, Inc. will switch from AT&T to **TINKERBELL** for all travel credit card calls. Employees traveling on BAP business are eligible to receive a company **TINKER-BELL** credit card. All AT&T credit cards should be returned to Finance & Accounting (3rd floor) by C.O.B. Friday, May 7, 1991. We will close the AT&T account at that time.

The **TINKERBELL** system requires a different procedure for using the credit card:

1. Dial the **TINKERBELL** Travel Number, 1-800-555-4311: you will get a voice prompt asking for your "code."
2. Enter your four-digit authorization code, then your two-digit travel code, which are printed on your card.
3. Enter the area code and the telephone number that you want to reach. These instructions are also printed on the front of your credit card.

Each business group has its own four-digit authorization code and travel code. Calls will be charged against your group's overhead. Finance and Accounting will provide group managers with itemized call sheets so they can track their long distance credit card expenses. Separate cards can be issued for specific contracts if necessary.

I am your principal contact for the new credit card system. I will assign authorization codes and travel codes. If you have problems dialing a number and have questions about procedures, contact me. If your credit card is lost or stolen, contact me or **TINKERBELL**, Inc. immediately.

Answers to Exercises

(From p.66) Final exercise: Apply verbal and visual coherence devices to help readers skim, follow, and refer back to the following paper.

Merged Account Legislation

This paper describes fixed accounts prior to the enaction of the law, how the law changed fixed accounts, and the effect of the law on BAP. Public Law 101-510, enacted November 1990, changed accounting and reporting procedures for **Merged** accounts. The legislation eliminated **Merged** accounts, and replaced them with a revised definition of **Expired** accounts and the newly created **Closed** accounts.

Prior to Public Law 101-510, fixed accounts, which we in BAP call annual and multi-year appropriations, had three states: **Active**, **Expired**, or **Merged.**

- **Active** accounts still had obligational authority. New obligations might be established during this period.

- **Expired** accounts started and lasted for two years past the expiration of an **Active** account's obligational authority. Obligation adjustments might be made during this period, but no new obligation might be recorded. When an an **Active** account expired, the unobligated balances returned to the Treasury surplus fund.

- **Merged** accounts began with the third year after the point of time when the obligational authority of the account expired. At this time, the obligations were grouped with all previous budget fiscal years, and the unobligated balances were grouped at Treasury to form the merged surplus fund authority.

After Public Law 101-510, fixed accounts have three states: **Active**, **Expired**, and **Closed**. Each state has a different meaning.

- **Active** accounts stay the same. **Active** accounts still have obligational authority.

- **Expired** accounts now have a *five-year* period following the expiration of obligational authority. (Previously, **Expired** accounts lasted only two years.) **Expired** account unobligated balances do not return to the Treasury as surplus funds, but remain with the obligated balances. During this period, all funds remain for recording, adjusting, and liquidating any obligations properly chargeable to the account prior to the time balances expired. Therefore, prior year adjustments can be made.

- **Closed** accounts now replace **Merged** accounts. **Closed** accounts, known as canceled accounts, begin with the sixth year after the obligational authority expires. Then, obligated and unobligated balances return to the Treasury as surplus funds.

Step 9. Edit for Clarity

(From p. 69) Exercise: Circle the abstract and general words (here shown in italics). Suggest concrete and specific words. Use your imagination.

Our office has been in *communication* with *your office recently* regarding a *question* about issuing a *refund for the remaining portion* of your lease on *some equipment*. There are *several reasons* why a refund of the *whole remaining portion* fails to satisfy *various material conditions* of the lease. However, your office may *contact our department* to negotiate *some amount* suitable.

BAP's accounting office wrote Ace. Co. last Tuesday, March 8, regarding your request for a $300 refund for the last two months remaining on your photocopier lease. Two material conditions of the lease preclude a full refund. You may contact Ms. Cook at our Finance Department, (800) 555-4561, to negotiate a lease buy-back price.

(From p.69) Exercise: Write two concrete or specific words for each abstract or general word.

1. transportation	train	airplane
2. subsidy	$10,000 scholarship	food stamps
3. fast	85 mph	2 copies per second
4. contact	telephone	meet
5. consider	interview for job	study for two weeks
6. move	lift	push
7. benefit (noun)	health insurance	401K pension plan
8. familiarize	learn	overview
9. various	three	6 point to 24 point fonts
10. response	letter	raised eyebrow

Answers to Exercises

(From p.71) Exercise: Identify each italicized verb as active or passive. (Answers show passive voice in italics, active voice in bold.) Convert passive verbs to active. Add a subject by answering the question *by what?* or *by whom?*

1. *It is pointed out* in the article that 13-column spreadsheets **help** bankers organize their information.
The article **points** out that 13-column spreadsheets **help** bankers organize their information.

2. Errors that I **have been making** for years *are* now more easily *seen* when I **edit.**
Errors that I **have been making** for years I now **see** more easily when I **edit.**

3. Minor errors can *be eliminated* through careful editing.
You **can eliminate** errors through careful editing.

4. The writing *was done* by a team of experts.
A team of experts **did** the writing.

5. Little attention *is being paid* to that advertising.
The public **pays** little attention to that advertising.

6. The client *was invited* by us to review the proposal.
We **invited** the client to review the proposal.

7. The verification and validation tests *will be conducted* after the terabyte of Landsat data *is loaded* into the database.
BAP **will conduct** the verification and validation tests after NASA **loads** the terabyte of Landsat data into the database.

8. The insurance investigation *is started* only after a legal complaint *has been submitted.*
Start your insurance investigation only after the client **submits** a legal complaint.

9. After the contract *was won*, we **met** the client to determine how the deliverables *would be accepted.*
After we **won** the contract, we **met** the client to determine how the client would **accept** deliverables.

10. Sorry—your money *cannot be refunded.*
Sorry—we **won't refund** your money.

(From p.73) Exercise: Put all verbs in present tense.

User Manual for Lawnmower

Follow these instructions to operate your lawnmower:

1. First, check the oil and gas levels.

2. Then ensure no debris is near the blades when you start the motor.

3. Next, put the choke to the red line. (See figure 2.)

4. Grab the deadman lever with one hand, then pull the starting rope with the other.

5. If your mower starts with difficulty, prime the carburetor.

6. After you mow your lawn, clean grass cuttings from the engine area.

(From p.73) Exercise: Put all verbs in present or past tense.

We have a copy of your resume in which you say you are able to begin work in September. I forwarded your resume to our Pawley's Island project officer, Edwina Smith. I also sent her a copy of the application you filled out when you came for your initial interview.

You can expect the following sequence of events. Edwina Smith returns in two weeks from a bidders conference held this week. Then, she calls you, and you and Edwina decide if you two conduct your second interview at the Pawley's Island site. I think that you match her requirement for a hydrologist, and that you may enjoy the Gulf Stream Project. We pay for travel expenses.

I look forward to hearing if you take the hydrologist position at Pawley's Island. Call me in three weeks if you do not hear from Edwina.

(From p.75) Exercise: Change subjunctive mood to indicative.
Dear Joe Palmer:

This letter clears up a misunderstanding about proposed changes to our purchasing policy, which affect your department.

Apparently, you took our first draft guidance too literally. You had an early draft. [Note that we cut completely the hypothetical statements: *should have warned, could have saved.*] Whereas the first draft told vendors not to expect to be paid in less than 600 days, the final policy tells vendors not to expect to be paid in less than 60 days.

Note other changes in the final draft of our purchasing policy. Please address further questions about the new vendor policy to Mr. Smith, (991) 555-1234.

Change these subjunctive sentences to make them clearly conditional.

1. Consider buying a house if you get a raise.

2. If an employee causes a late report, take disciplinary action.

Make these polite but ambiguous requests clear.

1. Please do not feed the animals.

2. Do not use flash photography in the planetarium.

Answers to Exercises

(From p.77) Exercise: Circle the ambiguous pronouns and suggest changes. (We put the changes in italics.)

Pennsylvania Department of Transportation Letter:

Dear Mr. Sparge,

For a road to be accepted into the Secondary System of State highways, *the road* must meet two criteria.

First, *the road* must serve three or more separate households. This *criteria* is frequently referred to, somewhat misleadingly, as the "three driveways" requirement. If your local road does indeed provide exclusive service to three households, *your road* is eligible for addition to the system.

Second, the road must meet state construction standards and must be in good condition. If work needs to be done for the road to meet this *criteria*, *the road* must be financed by either the residents or the original developer before *the road* is admitted as a standard subdivision street.

If state funds are needed, *you might use* the Rural Addition process. This process allows substandard roads to be taken into the system and public money used *for repairs*. Frequently, a special tax assessment is levied on residents to assist in paying for *these repairs*. Your local road may qualify for consideration under the Rural Addition process.

If you want more up-to-minute information on this *Rural Addition process*, please contact me at (800) 555-1234. Thank you again for bringing this *situation* to my attention.

From a specifications document:

The learning curve for the Foxglove report generator may take up to a week. If the application has simple report requirements, *you may find hard coding the report more efficient.* However, after *you* become familiar with Foxglove, *you* find *Foxglove* easy to include in applications.

From the TDR Queue window, choose the desired item from the list box by clicking on *the icon*. *This step* highlights the item.

(From p.79) Exercise: Match non-English or non-standard phrases with one of the standard English phrases listed on the right.

1. et al	g. and others
2. et cetera	r. and so forth
3. e.g.	b. for example
4. i.e.	j. that is
5. ergo	e. consequently
6. ad hoc	a. to this (purpose)
7. vice versa	n. position reversed
8. laissez faire	s. let people do
9. a la mode	o. in the fashion
10. in situ	k. on site
11. via	i. by way of (not means of)
12. vis-a-vis	q. face-to-face
13. ipso facto	c. by the fact itself
14. per	h. according to
15. in lieu	p. instead of
16. per diem	f. by the day
17. non sequitur	t. it does not follow
18. apropos	d. appropriate
19. departmentality	m. noun for departments?
20. conceptwise	l. adverb for concept?

Answers to Exercises

(From p.81) Exercise: Change these negative statements to positive statements.

1. Your problem is not justifying the need for a company car, but rather obtaining the necessary funding.
2. I find everything sobering in those dealings.
3. The Court upheld a state law allowing the investment tax credit.
4. Speak up.
5. The operator must remove the red safety tag from the disk drive to boot up the system and continue software installation.
6. The General Partners are liable for non-performance if the Limited partners prove that the General Partners acted in bad faith.
7. We rejected your bid because we accepted a less costly bid for the services.
8. Employers often give employees bonuses at Christmas.
9. I intend to appear reasonable.
10. Her response was logical, but it lacked key information.

(From p.83) Exercise: Think of a suitable synonym.

1. businessman	professional
2. craftsmanship	skill
3. foreman	supervisor
4. middleman	broker
5. sportsmanship	fair play
6. stewardess	flight attendant
7. fatherland	homeland
8. gentlemen's agreement	informal agreement
9. salesman	sales representative
10. waitress	server

(From p.83) Exercise: Edit these sentences to avoid gender bias.

1. Experienced ~~waiters~~ *servers* make dining more pleasant.

2. The average American drives ~~his~~ *a* car every day.

3. A ~~man~~ *person* who wants to get ahead works hard.

4. If ~~a man~~ *you* plan ahead, ~~he~~ *you* can retire at age 60.

5. Each senator selects his *or her* staff.

6. Be sure to bring your ~~husband~~ *spouse* to the D.C. Armory Flower Show.

7. Reagan, Gorbachev, and ~~Mrs.~~ Thatcher dominated politics in the 1980s.

8. The ~~fireman~~ firefighter and ~~policeman~~ police officer controlled the crowd.

9. ~~A~~ *Homeowners* can deduct interest expenses from ~~his~~ *their* taxes.

10. ~~The user~~ *You* can make only three attempts to enter ~~his~~ *your* password before the machine locks ~~him~~ *you* out.

(From p.85) Exercise: Make these sentences parallel.

1. The choice between an optimum system design or *a less desirable* one is affected by our R&D budget and *by* how we use commercially available software.

2. Management *assesses* your job performance by the following criteria. Are you neat and well-groomed, *do* you get your assignments done on time, *are* you flexible, and are you willing to learn.

3. Our latest magazine issue will lose money because we did not fill the advertising space, we needed 2,000 extra copies for promotion, and we *paid* too much for paper.

4. According to the request category, we will ~~either~~ recommend *either* the zoning board approve the plan outright or the review committee request more information.

5. When you make the list, arrange the items in order of importance, write them in parallel form, and *number* all the items.

6. We propose the following agenda for the meeting:
      a. *Call* the meeting to order.
      b. Set date for next meeting.
      c. *Take* the roll call.
      d. *Elect* new officers.

7. The tax committee voted to
      review the materials being purchased for the tax library
      submit a report on new billing rates
      develop client programs
      plan annual tax department party

8. By next Monday, please complete the research, analyze the various positions, and ~~you should~~ hand in the report.

9. The new accounting software package fails to meet our requirements for several reasons:
      1. It is too slow.
      2. *The menus are too* complicated.
      3. It *loses* all my data *if* power fails.
      4. The ledger *does* not balance *if* you enter future date.

10. When you build your database, ~~either~~ use *either* dBase IV for Windows or Altbase for OS/9.

Answers to Exercises

(From p.87) Exercise: Correct misplaced modifiers.

1. Our department receives *only* a limited amount of money to spend on office equipment.

2. The notices for the employees' benefits meetings were all published in the employee newsletter ~~along with~~ and in e-mail sent to our field offices.

3. The Fish and Game Club announced tuna *off the west coast* are biting.

4. I allowed the staff to take a day off *without pay* before I took my vacation.

5. A recent White House report *by the President's scientific advisor* was released claiming that acid rain is linked to methane emissions.

Or— A recent White House report was released *by the President's scientific advisor* claiming that acid rain is linked to methane emissions.

6. I have enclosed our company's financial statements *for your information.*

7. Senior managers will meet *in the board room* with the Chairman and CEO at 10:00 Tuesday morning about the Competition.

8. We watched the space shuttle *soaring high above the clouds* fly into space.

9. The Department of Housing receives *only* limited funds for maintenance.

10. Two campers were found *by the park rangers* shot to death.

11. I am happy to report, 18 districts reported no lost-time accidents *in April.*

12. NL Inc.'s goal is to report ~~any~~ *no* accidents, so it is extremely important to reduce and eliminate lost-time accidents.

13. I met a man *named Smith* with a wooden leg.

14. Please mark your calendars for the annual tax meeting *on April 5.*

(From p.87) Exercise: Correct the dangling modifiers.

1. To determine the final costs, *you* total the man-hours and multiply by the hourly rate.

2. After *the ship lay* on the bottom of the Atlantic Ocean for seventy years, the photographers brought back pictures of the Titanic.

3. Having studied the client's requirements, *we conclude that* the technical approach must include icon-driven menus.

4. Our user manual can not satisfy our Korean customer *who does not read English.*

5. To be a successful manager, *you need* good writing skills.

6. After *I wrote* a sentence outline, the first draft was easy.

7. Confident of our success, *we sent* the proposal to the client.

8. After *it rained* all day, we moved the reception indoors.

(From p.88) Final exercise: Abstract and general words, passive voice, shifting tenses, ambiguous pronouns, Latin words, negative statements, gender bias, an unparallel vertical list, and a misplaced modifier make this passage unclear. Edit for clarity.

Mr. Smith
P. O. Box 2456
Arlington, VA 22145

Dear Mr. Smith:

We reject your request to register your barn as a historical building for four reasons:

     1. Although the barn was built in 1927, by an alleged descendent of Lord Fairfax, the Historical Society places the barn's effective date much later because of improvements to accommodate cows.

     2. The barn requires extensive repairs that the owner and the Society can't afford.

     3. If the owner repairs the barn, the owner still must pay for insurance.

     4. The barn is already condemned to allow construction of a two-lane roadway.

Therefore, you cannot register your old cow barn as a historical site and impede construction of the new Route 230. Frankly, if we had not condemned the barn for road construction, your county intended to condemn your barn as a structural hazard. In that case, you—not the state—pay for razing your barn. Remember, we designed Route 230 to provide faster and safer transport of the many dairy vehicles.

Note how you can use the eight editing techniques for clear words and sentences:

1. Use concrete and specific words.
   for several reasons becomes *for four reasons.*

2. Make verbs active voice and present tense.
   have caused the Historical Society to place becomes *the Historical Society places*

3. Identify and replace ambiguous pronouns.
   it was designed becomes *we designed Route 230*

4. Use standard English words.
   Ergo becomes *In that case*

5. Be positive.
   We are not able to approve becomes *We reject*

6. Remove gender bias.
   for his own insurance becomes *for insurance*

7. Make sentences parallel.
   Neither the owner can afford, nor the Society becomes *the owner and the Society can't afford*

8. Test modifiers.
   you razing the barn instead of the state (implies you may tear down the whole state) becomes *you—not the state—raze the barn*

Step 10. Edit for Economy

(From p.91) Exercise: Cut empty verbs.

1. The manager performed the task.

2. The supervisor's reports expounded new theories.

3. The vice president wrote this contract.

4. The doctor removed my tonsils.

5. The team, despite its best efforts in the development stage, had to delay the start-up.

6. The position of the director, as reported in last week's newsletter, and commented on by many, remained precarious.

7. We attempted to repair the motorcycle.

8. The board of directors decided to notify employees about this year's payraises.

9. The auditors determined that BAPCO complies with generally accepted accounting principles.

10. We contracted the system design to Blue Communications, Inc.

11. The clients raised the following questions.

12. When deciding, remember we must meet safety at all times.

(From p.93) Exercise: Write alternatives to these prepositional phrases.

1. Call *at about* 5 o'clock.	at
2. *In accordance with* company policy. . .	following
3. He wrote *with the purpose of. . .*	to
4. Submit your plan *for the purpose of. . .*	for
5. Put the phone *on top of* the desk.	on
6. She is *in the midst of* a big job.	in
7. *In spite of the fact that. . .*	although
8. He is an expert *in the area of* finance.	in
9. Go *in back of* the shed.	behind
10. We are *in receipt of. . .*	have
11. He worked *over and above. . .*	beyond
12. *In the interest of* safety. . .	for
13. *With regard to* your promotion. . .	regarding
14. Indicate *as to whether or not. . .*	if
15. Because *of the fact that. . .*	(delete *of the fact that*)
16. Because *of this reason. . .*	(delete *of this reason*)

(From p.93) Exercise: Cut unnecessary prepositions from the following sentences.

1. To meet test objectives, XYZ, Inc. uses the La Jolla Laboratory staff's expertise.

2. Dr. Roger's review of the committee's draft report waits for DOD's review.

3. To qualify for the local business tax exemption, lecture series ticket sales must match these requirements.

4. The project ensures that the Navy receives maximum return on data documentation.

5. When deciding, remember safety.

(From p.95) Exercise: Cut *who, which, that*, and *there* from these sentences.

1. Ann Jones, ~~who is~~ the leader in our contract negotiations, wants to meet you on *your usual* six o'clock air shuttle. ~~which you usually take.~~

2. Please select a desk ~~that is~~ more suitable to your work.

3. Work continues on the Vega Project, ~~which is~~ scheduled for completion next summer.

4. He added a *similar* requirement. ~~that was the same as ours.~~

5. The policy committee, ~~which is~~ composed of local elected officials from Clark County, chose not to include a request for more road salt in their final *submitted* budget ~~that was~~ .

6. Remove the red safety tag, ~~which you'll find~~ next to the oil drain plug.

7. I hope ~~that~~ this letter answers your questions.

8. *We must discuss* ~~There are~~ three topics in our meeting.

9. If the customer requests statement copies ~~that are~~ older than six months, you must look in the microfilm library.

10. Employees must report any plant accident resulting in lost labor time to their shift supervisor ~~who is~~ responsible for safety.

(From p.97) Exercise: Cut the repetition.

1. He added a similar requirement.

2. Downsizing, as exemplified above, is key to the minicomputer boom.

3. Each stock item record contains a stocking conversion factor, the number of end-use units contained in one stocking unit.

4. This regulation is more important than others.

Answers to Exercises

(From p.97) Exercise: Cut unnecessary repetition from this passage.

Training Conferences. We planned three training conferences for government employees. The first occurs approximately 45 days after the contract award. We start with a working meeting and review of initial planning documents and requirements documents for training. The second training conference happens at day 90 to coincide with our first deliverable, the draft AIS training and technical manuals. We review customer comments of the manuals as well as the skills analysis report, plan of instruction, and course outlines. We also resolve concerns before we design and develop the training courses. The third conference happens about 225 days after contract award for the review and comment of the training materials and schedules, and for addressing other concerns. (Cut from 139 to 115 words or 17%.)

(From p.99) Exercise: Cut redundant words and phrases.

1. The subcontractor's concerns appear ~~to be valid and~~ important.

2. The determination ~~of whether or not~~ *if* standard~~, that is regulated,~~ procedures are required to manage ~~a series of multiple~~ performance tests is always subject to question ~~and is not susceptible to a conclusive determination.~~

Or—We don't know if we need standard procedures to manage performance tests.

3. BAMCorp's ~~singularly unique~~ personnel and technical package will ~~completely and professionally fulfill, as well as~~ satisfy~~, all~~ your ~~complex and challenging~~ requirements.

4. Doubling can ~~detract from and~~ confuse the message ~~or idea~~.

5. The company ~~especially~~ wishes to recognize ~~and compliment~~ Mr. Smith for five years of ~~unselfish and~~ generous ~~aid and~~ support to the Little League Baseball program in Falls Church.

6. The clients asked these questions ~~in their request for information~~.

7. From the airport #1: We want to be the first to wish you a ~~happy and~~ pleasant day in ~~the~~ Washington, D.C. ~~area or, if you are continuing your travels, we want to wish you~~ a pleasant trip to your ~~final~~ destination, ~~wherever that might be~~.

8. Also from the airport #2: Please ~~make sure~~ *secure* your tray-tables ~~are fastened and secure in an upright position~~ for landing.

9. From the airport #3: Please make sure your carry-on luggage ~~is of the type and size that~~ can be stored in the overhead bin ~~compartment~~ or ~~under the space~~ beneath the seat in front of you.

10. Customers can access their account information by modem, ~~that is, go on-line~~.

(From p.101) Exercise: Cut implied phrases.

1. ~~As you may already know,~~ Lockheed and Martin Marietta merged to become the world's largest defense company.

2. ~~It should be noted that~~ These new theories mark a radical change in the way scientists view the universe.

3. ~~All things considered,~~ Our new office manager shows promise.

4. ~~It is suggested that you~~ Send an invoice within 30 days of completing work.

5. ~~Please feel free to~~ Call me if you have ~~any~~ questions.

6. ~~When you find time,~~ Please ~~give~~ *tell* me ~~your decision about whether or not~~ *if* you want me to work late.

7. ~~Most experts claim that~~ Children need to eat a well-balanced breakfast before going to school.

8. ~~Before we begin our discussion,~~ Remember that these remarks are ~~strictly~~ off the record.

9. ~~We at~~ BAP Industries ~~would like to take this opportunity to~~ *thanks* ~~all~~ our vendors for their support ~~in our on-going activities~~.

10. ~~You may~~ Establish another category ~~for the purposes of~~ *to* record~~ing~~, adjust~~ing,~~ and liquidat~~ing~~e other obligations ~~properly~~ chargeable to the AIS contract.

11. Thank you ~~for your cooperation~~.

12. On the report ~~in question~~, please write the claim number in the blank space ~~available~~.

Answers to Exercises

(From p.103) Exercise: Cut unnecessary and vague modifiers.

1. I usually write my first drafts ~~very~~ quickly.

2. On the airplane: "Please use ~~extreme~~ caution when removing carry-on luggage from the over-head bins."

3. The clients asked the following ~~specific~~ questions.

4. ~~Actual~~ design of the system was contracted to Blue Communications, Inc.
Or—We contracted system design to Blue Communications, Inc.

5. BAP Industries offers a ~~most~~ unique solution to your ~~complete~~ personal computer needs.

6. Are you ~~absolutely~~ sure you unplugged the coffee pot?

7. Management remains ~~fairly~~ optimistic that we can meet our ~~relatively~~ high sales quotas.

8. We ~~greatly~~ appreciate your ~~outstanding~~ support.

9. Martin's analysis was ~~completely~~ accurate, but his conclusion was ~~totally~~ wrong.

10. Sally ~~first~~ debuted her ~~new~~ innovation to the public last month.

11. Write ~~up~~ the meeting notes, then pass them ~~out~~ to the committee members.

12. The inspectors were ~~rather~~ reluctant to sit ~~down~~ during the testing while I stood ~~up~~.

13. If you two will ~~both~~ cooperate ~~with each other~~, we can ~~all~~ achieve our goals.

14. Pam and John find it ~~mutually~~ agreeable to join ~~together~~ in the bonds of marriage.

15. The auditors determined that BAPCO complies with generally accepted accounting principles. (Note: We usually cut empty adverbs like *generally*, but here the phrase *generally accepted accounting principles* or GAAP has special meaning for accountants. In this context *generally* is not deadwood.)

(From p.104) Final exercise: Try to cut 25-50% of the 302 words in the following passage. (This improved copy has 142 words, a 53% reduction.)

MEMORANDUM

TO: Staff
FROM: D . Rose
SUBJECT: Budget Planning Sessions

You already have the pages attached to this memo. They reflect our revision to the budget planning sessions schedule issued January 5. We changed the session dates and times to include current data from our field operations. Reorganization of field offices changed the budget projections for cost and revenues. Additional budgetary sessions may occur to review cost and revenue projections researched by our field offices.

Gather and prepare budget numbers before the first session. Use the attached budget format as a guide. We intend to cover each topic in detail.

If you desire to attend a budget session, and are not a field office budget team member, please tell me or any Finance Department team member so we can tell you of any schedule variances. Except Wednesday, each session meets 9:00 A.M. to 12:00 P.M. and from 1:30 P.M. to 4:30 P.M.

Note how you can use the seven techniques to cut deadwood.

1. Cut empty verbs.
   You should already be in receipt of becomes *You have*

2. Cut unnecessary prepositions.
   schedule of budget planning sessions becomes *budget planning sessions schedule*

3. Cut *who, which, that,* and *there.*
   schedule which was issued becomes *schedule issued*

4. Cut repetition.
   each daily session is scheduled to be an "all day" session meeting becomes *each session meets*

5. Cut redundancy.
   attached in the enclosure becomes *attached*

6. Cut implied phrases.
   It was felt that the recent reorganization becomes *Reorganization*

7. Cut unnecessary or vague modifiers.
   first and only schedule becomes *schedule*

Step 11. Check for Readability

(From p.107) Exercise: Measure readability by calculating the average sentence length and percent of long words for each passage.

Managing Proposal Commitments
Average Sentence Length = 31   % Long words = 35
How to Manage Proposal Promises
Average Sentence Length = 13   % Long Words = 9
Note: The second passage, which provides the same information as the first, is much easier to read.

(From p.109) Exercise: Replace these long words with short, one-syllable words.

1. accurate	right	26. initiate	start
2. actuate	move, start	27. magnitude	size
3. additional	more	28. methodology	way
4. allocate	put, place, mark	29. minimum	least
5. aggregate	group	30. modification	change
6. apparent	clear, plain	31. necessitate	need, force
7. ascertain	learn	32. objective	goal
8. assimilate	merge	33. operate	use, run
9. assistance	help	34. optimum	best
10. capability	skill	35. preliminary	first
11. commence	start	36. prioritize	rank
12. constitutes	is	37. probability	chance
13. demonstrate	show	38. remuneration	pay
14. denominate	name	39. represents	shows, is
15. designate	name	40. self-conscious	shy
16. disseminate	spread	41. sensible	wise, sane
17. eliminate	cut	42. stratagem	plan
18. enumerate	count	43. substantiate	prove
19. establish	start, place	44. suitable	right
20. expeditious	quick	45. terminate	end, kill
21. expertise	skill	46. uncompromising	firm
22. facilitate	help, ease	47. underutilize	waste
23. functionality	use	48. utilize	use
24. generate	cause, make	49. variance	change
25. identical	same	50. voluminous	huge, large

(From p.111) Exercise: Break these long sentences to improve readability. Use short sentences for emphasis, vertical lists to group related items, and long sentences to express complex relationships.

1. The client had told us that the tanker was purchased in December, 1986. After fulfilling an existing obligation to act as a storage facility for fuel in the Caribbean, the tanker proceeded to Portugal in May, 1987. *There* it was dry-docked for barnacle scraping, painting, and repairs.

2. ~~Because~~ The multi-state Rentacar discount is the only discount plan that would require these types of functionality. ~~and~~ No other discount plans have been proposed that might require this functionality. *Consequently*, the multiple levels or alternate level credits will not be addressed in any other functional specifications.

3. ~~To meet~~ The Air Force Controller needs a financial system that would provide a single, consolidated repository of a budget execution, general ledger, and external reporting for Air Force-wide financial management purposes. We developed

- MegaCount software modifications
- custom interface programs to provide the MegaCount application software with data from external Air Force budget execution and reporting applications
- conversion programs to convert existing Air Force data to MegaCount formats and data files
- additional custom reports, including external reports for submission to Treasury and GAO

4. The purpose of the Uniformed Securities Act is to protect investors from fraudulent securities transactions for which the administrating agency requires securities to be registered with the state. ~~and~~ Unless a security is specifically exempt from registration, or the transaction is considered exempt, the security must be registered before it can be sold, or offered for sale within the state.

(From p.112) Final exercise: The original passage *(believe it or not, an advertisement)* has 40% long words and an average sentence length of 48 words. If you got down to 10% long words and 20 words per sentence, congratulations. In our answer, we took the readability down a peg to show that you can ruthlessly cut the deadwood, replace long words, and cut sentence length without losing meaning.

UPS Saves Computers

Protect your company's computers. Reduce mean time to failure for CPUs. Improve hard disc speed. Increase the life of input/output devices. To protect your machines, you need an uninterruptible power supply (UPS). We suggest the Microman Standby System. Its internal EMI/RFI filters and surge protection protect your machines and data from power failures. These UPSs feature great power transfer rates, complete diagnostics, an LED status panel, smart interfaces, alarms, and handsome casings. End power-related risk.

Average Sentence Length = 9   % Long Words = 9

Notes:

1. To help cut empty verbs, unnecessary modifiers, unnecessary prepositions, and implied phrases, change (or cut) the general and abstract words. For example, what is *a pre-eminent prerequisite*? or *a technologically superior power transfer rate*? *a secure preventive maintenance environment*?

2. Eliminate unnecessary modifiers like *obviously, excellent, essential, serious, pre-eminent, marvelous, virtually*, and many others.

3. Change verbs to active voice, present tense, and imperative mood. The imperative mood cuts all the bulky references to *configuration manager.*

4. Cut redundancies. For example, Input/output devices are *peripherals.*

5. Cut implied phrases. Alarms are *audible.*

6. Replace three-or-more syllable words with one-or-two syllable words. For example, *performance* becomes speed. Power *interruptions* becomes power failures. *Eliminating* becomes end.

7. Cut compound sentences.

Step 12. Check for Correctness

(From p.115) Exercise: Define these commonly confused words. (The matched definitions are in bold. Other sound-alike words are on the right.)

1. *except*	to receive with favor	accept
	aside from	
2. *adapt*	highly skilled	adept
	to take as one's own	adopt
	to adjust to the situation	
3. *advice*	**a noun meaning counsel given**	
	a verb meaning to recommend	advise
4. *effect*	to produce a change	affect
	a result (noun)	
	to result in (verb)	
5. *ensure*	to promise someone	assure
	to make sure	
	to protect against loss	insure
6. *capital*	**seat of government**	
	money owned	
	the building where legislators meet	capitol
7. *cite*	**to use as proof**	
	to summon to appear in court	
	act of seeing	sight
	that which is seen	sight
	place or location	site
8. *compliment*	that which completes	complement
	a flattering comment	
9. *counsel*	a group of people	council
	advice (noun)	
	advise (verb)	
10. *discrete*	tactful	discreet
	separate or distinct	
11. *everyone*	**every person of a group**	
	every person	every one
12. *farther*	**space or distance**	
	to a greater degree	further
13. *formally*	**according to custom**	
	in the past	formerly
14. *it's*	a possessive pronoun	its
	contraction of it is	
15. *lie*	to place	lay
	to recline	

16. *lone*	the act of lending (verb)	loan
	that which is lent (noun)	loan
	by oneself	alone
	isolated	
17. *lose*	not fastened or confined	loose
	to part with	
18. *past*	moved on (verb)	passed
	at a former time (adjective)	
	former time (noun)	
19. *personal*	**private**	
	employees	personnel
20. *principal*	**leader**	
	money (noun)	
	first or highest (adjective)	
	rule	principle
21. *respectively*	showing respect	respectfully
	considered singly	
22. *write*	upright	right
	correct	right
	solemn act	rite
	to make words on a surface	
23. *stationary*	**standing still**	
	letter paper	stationery
24. *statue*	height or level	stature
	rule or law	statute
	carved figure	
25. *than*	**in comparison with**	
	at that time	then
26. *their*	**possessive pronoun**	
	at that place	there
	contraction of "they are"	they're
27. *too*	toward	to
	in addition	
	one more than one	two
28. *addition*	**increase**	
	attachment	
	publication	edition
29. *alter*	religious table	altar
	change	
30. *basis*	reasons, or foundations	bases
	a reason, or a foundation	
31. *biennial*	twice a year	semiannual
	every two years	

32. *devise*	equipment **plan**	device
33. *disapprove*	**have an unfavorable opinion** show to be false	disprove
34. *elicit*	**ask for** illegal	illicit
35. *illegible*	qualified **unreadable**	eligible
36. *eminent*	**prominent** about to happen	imminent
37. *envelop*	**surround** container for a letter	envelope
38. *expend*	increase **pay out**	expand
39. *physical*	financial **of material things**	fiscal
40. *forward*	preface **at the front**	foreword

(From p.119) Exercise: Circle the subject(s) and underline the verb(s) of each sentence. (Subjects are in bold; verbs are in italics.)

1. **Thunder** *is* loud, but **lightning** *does* all the work.

2. The **customer** *does* not *know* what **we** *can do* for her company.

3. **Fred and Barney** *took* Wilma and Betty dancing.

4. Your **contribution** to the project *deserves* our praise.

5. **Sticks and stones** *may break* my bones, but **words** *will* never *hurt* me.

6. **You** *can't win* if **you** *don't play.*

Answers to Exercises

(From p.119) Exercise: Identify each word group below as a correct sentence (C), incomplete thought (IT), run-on (RO), or comma splice (CS).

1. IT Considering that the competition has reacted strongly to our effort to grab more market share.

2. IT Can type twenty-five words per minute.

3. IT Mr. Johnson, unable to attend the afternoon meeting or evening dinner.

4. C Beverly Timmons, project leader for database development, made three unsuccessful requests for government assistance.

5. C Whose responsibility is it to clean up the oil spill?

6. IT Until we found out that Good Food Inc. had raised its price to cater a cocktail party and the Sheraton Inn had almost doubled the price to rent the ballroom.

7. IT Now that Sandra has finished her Associate Degree in Accounting.

8. IT Tax increases choking off economic growth again.

9. C The office manager interviews all candidates for staff positions.

10. IT Friendly, courteous, and always available to answer your questions about our software products.

11. C The favor of reply is requested.

12. CS We have a scheduling conflict for the conference room, Mr. Smith scheduled a news conference at four and the facility engineer planned to recarpet the floor, please advise.

13. RO The climb to the top is hard remember that staying at the top is harder.

14. CS Concentration is the key to economic success, it's also the key to success in life.

(From p.121) Exercise: Circle the correct pronoun. (Correct answer is in bold.)

1. The two winners were Jane Swanson and (**I**, me).

2. Please send Mr. Jenkins and (I, **me**) to the seminar.

3. Both you and (**he**, him) should apply for the new position.

4. The telephone technician (who, **whom**) you sent for has helped us before.

5. Rebecca is taller than (**I**, me).

6. No one wants to win the AIMS job more than Alice Cairnes and (me, **I**).

7. The company must monitor (**its**, their) sick leave policy carefully.

8. BAP Industries, Inc. has (**its**, their) headquarters in Virginia.

9. The team won (**its**, it's, their) first game of the season.

10. (Its, **It's**) not (**I**, me) (whose, who's, **who am**) responsible for losing the key!

11. (Their, **They're**) talking about (**your**, you're) book.

12. (Whose, **Who's**) in charge of marketing?

13. The committee can't agree what (**its**, it's, their) responsibilities are.—if the sense is the committee's collective responsibilities.

     The committee can't agree what (its, it's, **their**) responsibilities are.—if the sense is the committee members' individual responsibilities.

14. That's (**he**, him) standing in the lobby.

15. Jerry writes better than (**they**, them), so (**their**, they're, there) supervisor asked (he, **him**) to edit the company newsletter.

Answers to Exercises

(From p.123) Exercise: Circle the subject. (Answers are in bold.) Write the verb in the form that agrees with the subject. (Changes are in italics.) Some sentences are correct.

1. **Each** of the four divisions in the company is responsible for its own costs.

2. Neither the **Army** nor the **Air Force** wants to pull troops out of Germany.

3. The senior **scientist and engineer** in this company *want* to work on the space-station contract.

4. Difficult **decisions** like the one we must make today *take* time.

5. A **collection** of paintings by three local artists is on display in the lobby.

6. A four-member **crew** cleans and maintains each UPS truck.

7. Where *do* the **desk, chair, sofa, and filing cabinet** go?

8. **Shoes, belt, and a tie** add a lot to a man's wardrobe.

9. The **board** of directors *agrees* with management.

10. The **carton** of typewriter ribbons *is* sitting on the desk.

11. The **duties** of the police officer *require* courage and self-sacrifice.

12. **Attention** to details *ensures* fewer errors.

13. **Both** have the authority to write checks up to $1,000.

14. **George Burns**, with his companion Gracie Allen, needs no introduction.

15. Here *are* the new **copy machine and its instruction manual**.

16. **Half** a load of bricks *does* not satisfy our order.

17. I wish **I** *were* your boss instead of your assistant.—subjunctive form of "to be" = were.

18. If **wishes** were horses, then poor **men** would ride.

19. If **I** *were* a full time employee, **I** would get a salary with benefits, but **I** would lose my over-time.

20. **ACME Theaters** is a large national chain.

(From p.125) Exercise: Write *a* or *an* before each word.

1. a   10 percent raise   10. an   11 percent drop

2. an   action   11. a   balancing act

3. an   example   12. a   donor

4. a   European   13. an   FBI investigation

5. an   hour   14. a   history book

6. a   hostess   15. an   icon

7. an   M.B.A.   16. a   one-time write off

8. an   order   17. an   S.O.S.

9. an   uncle   18. a   uniform

(From p.125) Exercise: Write *a* or *an* in each blank.

1. A one-month night shift is followed by an eleven-day paid leave.

2. Dr. Peters, a history professor, taught a unit about the Civil War.

3. An ambassador from a European country made an unusual request at a UN meeting.

4. His clock has an electrical dial, not an hour, a minute, or a second hand.

5. An aspirin is not always enough for an aching head.

6. Jane earned an M.B.A. with an emphasis on marketing.

7. We need 100 days to complete an order, but we have an 83-day deadline.

8. The doctor told me to start an aerobic activity, which was not an answer I wanted to hear.

Answers to Exercises

(From p.125) Exercise: Correct the double negatives in these sentences.

1. James couldn't find ~~hardly~~ anyone to invest in his gourmet doughnut shop.
2. Nobody ~~doesn't~~ *wants* to miss the staff meeting.
3. Peter doesn't know *anything* about the value of a dollar.
4. If you don't have a positive attitude, you won't succeed. —OK
5. She couldn't ~~hardly~~ hope to get a promotion after one week on the job.
6. Wouldn't Jim rather ~~not~~ go? or Wouldn't Jim rather not go?
7. Let's ~~not~~ give no more thought to the unfortunate incident.
8. Phyllis never ~~hardly~~ saw ~~no~~ records to justify Bill's tax deductions.
9. Roger never met *anybody* that didn't ~~not~~ like his mom's toll house cookies.
10. The safety inspector told us that we must not store ~~none of~~ the nitro next to the glycerin.

(From p.127) Exercise: Write the singular possessive and the plural possessive.

1. clerk's clerks'	5. Jones's Joneses'	9. day's days'	13. business' businesses'	17. man's men's
2. facility's facilities'	6. area's areas'	10. boss's bosses'	14. knife's knives'	18. advisor's advisors'
3. fence's fences'	7. city's cities'	11. guest's guests'	15. line of credit's lines of credit's	19. loss's losses'
4. waitress' waitresses'	8. year's years'	12. lunch's lunches'	16. brother-in-law's brothers-in-law's	20. friend's friends'

(From p.127) Exercise: Insert an apostrophe and an *s* to show a possessive noun. Make other nouns plural if necessary. (Changes are in italics.)

1. The *employees* attend lectures where they learn techniques to improve their *plant's* efficiency.

2. We investigated *Bill's* complaint.

3. All of the *questions* were answered in turn.

4. The *Treasurer's* recommendation is that we cut overhead *costs*.

5. If we build a *men's* locker room, we'd better build a *women's* locker room too.

6. We hired several new *employees* for the Jason project.

7. *Fred's* office will need two *coats* of paint.

8. It's the office *manager's* responsibility to make sure the *lights* work in the conference room.

9. The toxic waste response team must respond in a *minute's* notice.

10. Please evaluate *Avis'* proposal to discount our *company's* large car rental fees.

(From p.129) Exercise: Insert commas where needed.

1. The Smith Foundry Tool and Die Company was started by Thomas Smith, an inventor and entrepreneur.

2. I want to see last quarter's income statement, balance sheet, and cash flow statement.

3. Margery, Elizabeth, Susan, and I can just barely fit in her new sedan.

4. IBM, Compaq, Apple, and a host of other personal computer manufacturers are struggling to define their marketing strategies.

5. Employee benefits include paid vacation, holidays, sick leave, bereavement leave, and unpaid maternity leave.

6. The new copy machine is faster, cleaner, more reliable, and more versatile.

7. We can buy industrial grade, fire retardant carpet in either beige, dark blue, green, gray, or burnt orange.

8. Our lounge always keeps pots of regular and decaffinated coffee, with cream and sugar.

9. Your flight has stops in Atlanta, Denver, Los Angeles, and Melbourne.

10. Because the air conditioner broke down, we're releasing the workers at 2:00 p.m.

11. After I just got a $2 million construction loan, you've got a lot of nerve telling me you underestimated the job.

12. Although they finished paving the parking lot, they have not painted the lines yet.

13. Gigamega, the most powerful computer ever built, has been programmed to invent video games.

14. Our new vice-president for engineering, Dr. Potts, will lead the discussion on cryogenics.

15. William's plan, even though it made no sense to us, won high praise from the Navy.

16. If I had to learn a second language, all things being equal, I would study FORTRAN.

17. Patriots Day, a paid holiday in Massachusetts, does not merit a day off in Virginia.

18. Slick Magazine, boasting a circulation of 5 million paid subscribers, charges $1600 for a quarter-page ad.

19. In the past, success came easily to George.

20. In conclusion, we use commas to separate parenthetical expressions from the main idea of the sentence.

(From p.131) Exercise: Insert semicolons where needed.

1. The receptionist area needs new carpet; however, we'll wait until we remodel the entire floor.

2. Although the receptionist area needs new carpet, we'll wait until we remodel the entire floor.—no changes.

3. Dr. Latrobe, a propulsion expert, designed a rocket motor that runs on normal jet fuel; nonetheless, liquid hydrogen remains our preferred fuel, because it has a better thrust to weight ratio.

4. Francis will meet us at O'Hare Airport, however, thirty minutes later than expected.—no changes.

5. Luck is where preparation meets opportunity; so keep your eyes open and be prepared.

6. We won the contract; now we have to do the work.

7. Karen Kelly brought us some of our most profitable accounts; for example, she landed both the Hechinger and the Safeway accounts.

8. We've added four new sales districts, which are Atlanta, Georgia; Mobile, Alabama; New Orleans, Louisiana; and Houston, Texas.

9. Our company has but one mission, that is, to provide our clients the best value in video home entertainment.—no changes.

10. Megatech bid the highest price; nevertheless, they won the contract on technical merit.

11. Conglomerator, Inc.'s most recent acquisitions were Catfish Farms, Ltd. on April 10, 1989; and Carlisle Cosmetics, Inc. on December 2, 1990.

12. Although Mr. Derickson is younger than the other applicants, he should get the job because of his superior performance record.—no changes.

13. Our company's policy is to promote from within; for instance, Mr. Jacobs started as a clerk and rose to be chief executive officer.

14. Because John Heath just came to us from the Department of Transportation, where he was a special assistant to the Secretary, we mustn't bid him on the Highway Study Project.—no changes.

15. As long as sales continue to increase at the present rate, we can absorb the rising cost of labor without raising our prices.—no changes.

(From p.133) Exercise: Insert colons where needed. Check capitalization.

1. Next time, give that pushy salesman an evasive answer: Tell him to take a long walk off a short pier!

2. Note well: The company would have posted a substantial loss in 1990 except for the one-time sale of the Occoquan property.

3. Our company has but one mission: to provide our clients the best value in video home entertainment.

4. Conglomerator, Inc.'s made two acquisitions: Catfish Farms, Ltd. on April 10, 1989; and Carlisle Cosmetics, Inc. on December 2, 1990.

5. You must add one procedure to lower your worker's compensation insurance: You must aggressively prosecute fraud.

6. Eric made a significant breakthrough in his research: He discovered a new graphite compound.

7. Doc Watson put a sign on his briefcase: Moon or Bust!

8. The odor from the paper mill smoke smells bad, but it's harmless.—no change.

9. Managing inter-personal conflict is like the law of thermal dynamics: You can't win, you can't break even, and you can't get out of the game.

10. The employee lounge has three simple rules for everyone's mutual enjoyment: no smoking, no alcoholic beverages, no radio-players without earphones.

(From p.135) Exercise: Insert dashes, parentheses, commas, or colons. Some sentences can be punctuated several ways.

1. The invoice for $215.00 (not $21.50) needs your prompt attention.

2. Writing and editing ability—that's what we want in our senior technical staff.

3. The partnership usually pays a portion of the net profits: for example, $670 per limited partner in 1990 to help cover the limited partners' tax liability.

4. Jeff must fix the rear projector in the conference room today—tomorrow is too late.

5. The Penultimate II cordless phone—you won't find a better value—offers the following features: speed dialing, auto call back, conference calling, and much more.

6. The offices on the the sixth floor (Treasury, Marketing, and Human Resources) will be moved to the new building in May.

7. Dorothy had it all: a dog, ruby slippers, and Kansas.

8. Rebecca requested a four-week vacation: she won a cruise, and she has no choice as to dates.

9. Sales rose (See figure 4, page 32) to a record high—however, return on sales fell slightly.

10. A high school diploma, three years' experience, good references—these are the minimum requirements.

Answers to Exercises

(From p.137) Exercise: Punctuate compound adjectives with hyphens.

1. We graduate a hundred-odd students each year.

2. John sent a carefully worded letter to IRS to explain his highly irregular filings for 1988 and 1989. —no change.

3. Send an up-to-date roster of security clearances to Lt. Avery.

4. The security inspector, Lt. Avery, said our security clearance roster was not up to date. —no change.

5. It's a pleasure to receive a well-written proposal.

6. He wished his off-the-record remarks had stayed off the record.

7. Alice made a reasonably good attempt to send the package before five o'clock.— no change.

8. John and Martha sublet a one-bedroom apartment in a not-so-nice part of town.

9. Thelma told the interior decorator she wanted egg-shell white paint in the dining facility, but when the paint dried, she swore the color was coffee-stain brown.

10. Mrs. Stern won't tolerate a gum-chewing receptionist.

(From p.139) Exercise: Correct the punctuation in these lists.

1. We need to address two conversion issues for Design Release 2.2. These are

      1) conversion of information from the VAX to IBM environment
      2) initialization and maintenance of the operator's manual

2. Avoid ambiguity:

      1) Choose words carefully.
      2) Place modifiers close to things they modify.
      3) Use active verbs and avoid passive voice.

3. Before you turn off the LAN host computer

      close any open files and exit any active programs
      run the backup to tape procedure
      run the LAN check to warn users of system shutdown
      input a valid LAN operator ID#
      shut down the LAN program

4. The strengths of the Shazbot system include

      a. error trapping prevents faulty data entry

      b. on-line help functions decrease training time

(From p.141) Exercise: Improve the mechanics of these sentences.

1. At our 9:00 A.M. meeting we reviewed the 4-month extension through September, 1990. We learned yesterday that Department of Defense (DOD) will request another 2-month extension. We decided to submit the extension to DOD for two months with two 1-month options.

2. Beginning March 19, 1995, Ben Brown will handle any material or supplies request through the MS/Plus computer system.

3. The mayor introduced former president Bush at the local Veteran's Day celebration in Alexandria, VA.

4. Enter your user-ID in the IBM; then transmit your Lotus files from the PC's hard-disk to your own floppy disk.

5. Two hundred and fifty people attended the Air Force convention in Palm Springs, CA.

6. Mr. Smith called this morning. (He left his telephone and fax numbers.)

7. The order was for twelve 6-inch pipes; we shipped them to Joe's Hardware Inc. yesterday at 5 P.M.

8. See figure 4 on page 9.

9. My favorite book is How to Repair Your Volkswagen: A Step-by-Step Manual for the Complete Idiot.

10. This occurrence only strengthens our commitment to proceed with BAP Industries' expansion into nickel mining.

(From p.142) Final exercise 1: Circle the correct word. (Answers are in bold.)

1. We are (adapt, **adept**) at software design.

2. Please indicate your (ascent, **assent**) by signing the contract.

3. The improved lighting has had a good (affect, **effect**) on productivity.

4. Careful pre-writing (assures, **ensures**, insures) effective writing.

5. Dewey, Cheetham and Howe serves as (council, **counsel**) to the city (**council,** counsel).

6. The salad makes a fine (compliment, **complement**) to the broiled fish.

7. It helps to break the problem into (discreet, **discrete**) topics.

8. (**Everyone**, Every one) must attend the safety briefing.

9. We cannot discount our hourly rates any (farther, **further**).

10. Can we (forego, **forgo**) the interview process?

11. Mr. Smith was (formally, **formerly**) self-employed.

12. The operator must get (**past**, passed) the shut-off valve before seeing the display.

13. We will not allow an employee administrative leave unless we believe there is a serious (personnel, **personal**) problem.

14. The key to persuasive writing is seeing the reader's (**perspective**, prospective).

15. The students felt sure of success because they had a (principal, **principle**) at stake.

16. Please examine each cost item (respectfully, **respectively**).

17. Please (sit, **set**) yourself a place at the table.

18. What harm is a couple of beers (between, **among**) friends?

19. The Fairfax County Symphony gave a (credible, **creditable**) performance.

20. Doctors recommend we eat a (healthy, **healthful**) breakfast.

21. There are (less, **fewer**) than six days left to complete the work.

22. We moved our offices to the suburbs (since, **because**, due to) the lease expired and (since, **because of**, due to) the high prices in the city.

23. As the company's founder, Mr. Adam Smith raised the company to a (**respectable**, respectful) position in the steel industry.

24. Refer to the letter, (that, **which**) I sent last Tuesday.—Note the comma after letter.

25. The lawyer had no (**further**, farther) questions for the witness.

(From p.143) Final exercise 2: Punctuate these sentences correctly.

1. If ever you've nothing to do, and plenty of time to do it in, why don't you come up and see me.—Mae West in the movie "My Little Chickadee."

2. In his best selling book <u>Wabbit Hunting</u>, the author Elmer Fudd discusses a hundred ways to trap, snare, or shoot cwazy wabbits.

3. Our bookkeeper, Teresa, impressed the auditors with her accurate files.

4. The temporary services agency Temps & Co. will give us eight hours of temp services at no charge, just so we can evaluate their company.

5. Erica, the company expert on time management, suggests that we conduct all staff meetings standing up.

6. We billed four hours at the principal rate of $180 per hour, and eight hours at the staff rate of $42 per hour.

7. Lorna Ewald, Ph.D. in computer sciences, started her own company in 1985, but she sold her interest to Logicon Inc., and then she came to work for us.

8. Population growth in the United States, according to the latest census data, has fallen if you take out immigration.

9. The qualities we seek include good people skills, willingness to learn, and willingness to travel.

10. Sam's motto—cash is king—makes a lot of sense in the 90s, when so many companies struggle under debt.

11. Steel, oil, and railroads—the great monopolies of the 19th Century—changed the face of capitalism forever.

12. Dr. Nathan, our only nuclear engineer, decided that the company's research into cold fusion is a poor investment.

13. A typical engine-overhaul is a one-day job.

14. Red, white, and blue will wrap our Fourth of July Sale in the flag.

15. The telecommunications van must be able to operate in the tropics; therefore, we added a dehumidifier to its on-board equipment.

16. Dr. Harold Brown, Chairman of the Loudon Board of Trade, met with the Loudon County Zoning Commission to attempt a compromise between local environmentalists and developers.

17. AMTRAK's Metroliner runs between Washington, D.C. and New York in 2 hours and 52 minutes, with stops in New Carrolton, Maryland; Baltimore, Maryland; Wilmington, Delaware; and Philadelphia, Pennsylvania.

18. Any member of our cross-trained staff (that would include me) can help you solve your most difficult files management problems.

(From p.144) Final exercise 3: Find and correct the word choice, grammar, punctuation, and mechanics errors in this excerpt. (Changes are in italics.)

Forecast Economics *prides* itself *on* the results *its* clients *receive* from *its* many time-tested econometric *models.* The success of a model depends on the assumptions upon which *it is* based. To *ensure* the accuracy of the assumptions it uses, *Forecast Economics tests* several *designs* to find the sensitivities of shifts in the assumptions. Moreover, Forecast *researches* brokerage costs, *price entry, and exchange fees. Its* research into specific market securities *has* also addressed the risks of volatility, or any other risks that come to mind.

Completeness and accuracy are the *goals.* For example*, Dr.* Brown will incorporate into the model any new regulatory information or new market theory *whenever* the thought occurs to *him. This* flexible and rapid response capability couldn't be replicated *anywhere* else. In the past, he *determined and quantified* the extent that the regulatory aftermath of the *October 19, 1987,* stock market crash *had* on the *Standard and Poor*'s Stock Index. Having successfully incorporated *this* kind of market aberration into econometric modeling, *Forecast Economics provides reliable real-time data modeling to its clients.*

Step 13. Proofread

(From p.147) Exercise: Proofread the following letter. Identify errors as word choice, grammar, punctuation, mechanics, typing, or content.

12 June, 1992

Graphics Leasing Corporation
Attn: Accounts Receivable Manager
VGS Park, Dept. A
5701 9th Ave., N.W.
Washington, DC 20005

Dear Accounts Receivable Manager:

Below is a list of Alcor Corporation check numbers, invoice dates, and amounts that are in payment of the lease and maintenance agreement on the POS One-320 camera presently in our possession. This information was obtained through our disbursement summary report for fiscal year 1989.

Because of the age of these items, neither canceled checks nor duplicate copies can be obtained for review. I suggest that you check your deposit records for the time in question to verify the amounts.

Check Number	Invoice Date	Amount paid
037614	4/25/89	$ 932.04
043569	4/25/89	804.12
012847	4/28/89	804.04
019383	5/14/89	425.12
036000	7/3/89	445.12
036001	8/15/89	388.24
037455	1/11/89	432.17
045672	11/22/89	464.25

Total Paid: $5,120.22

These problems have occurred because of your inaccurate files. The discrepancy is two and one half years old; therefore, we are not responsible for proving any further the payment of these items. I consider this matter closed with this letter and the schedule of paid items.

Sincerely,

Jane Walters
Accounting Supervisor

(From p.149) Exercise: Check live copy against the dead copy. What errors were corrected? What errors remain? Did new errors creep in?

Line	Error:
3	remains, change "your" to "you"
14	corrected, *"Stated"* to *"Started"*
17	new error, "men" to "menues" instead of "menus"
19	corrected, "user's" to "users"
24	corrected, delete period end of list
28	new error, deleted comma but also closed space
36-37	remains, change the double "you you" to "you"

(From p.150) Final exercise: Proofread this memorandum.

INTEROFFICE CORRESPONDENCE

To: Distribution
From: Mary Poole
Date: September 2, 1991
Re: Briefing for Training Facilitators

You have been assigned training responsibilities in conjunction with this year's MGR training effort. The two short tapes that will be used for the1992 training will be "Timekeeping" and "Harassment." The tapes are already in production and will be shipped to you approximately October 1, 1991.

A briefing session will be held at 11 A.M. on October 28, 1991, at headquarters to review the content of the tapes, identify expected discussion topics, and provide some direction for leading the discussion periods. I look forward to seeing you at the meeting. Although we may extend into the noon hour, you may expect that we will be finished by 2 P.M.

You should move forward with scheduling training sessions at your employees' locations, with the goal of completing all training by the end of August, 1992.

Distribution:
President
Comptroller's Office
All Senior Technical Staff

Answers to Exercises

(From p.150) *Final* Final exercise:

Now that you know the Writing System, use it to your advantage to

1. **Manage your time.** You can set realistic deadlines and track your progress. If you hit a snag and slip from your schedule, you can get help or notify others of the delay.

2. **Write faster.** With the Writing System and its techniques, you can focus your efforts; analyze and solve problems; sprint through your first draft; and quickly revise, edit, and proofread your document.

3. **Coordinate with others.** Use the purpose statement, sentence outline, Content Test, and Organization Test to get concurrence from team members, supervisors, or clients. Use the Pre-writing Phase to "flesh" out details and force controversies before writing the draft.

4. **Improve document quality.** The Writing System reduces the number of errors that creep into a document during edits. Moreover, the mechanical editing techniques are fast and accurate. The techniques force you to clarify vague words—especially pronouns. And you must state who or what performs the action. Checking for correctness and proofreading puts the final gloss on your well-written document.

As you practice the Writing System, you discover other benefits, which include

1. **Read more critically.** In time, you apply the Content Test to everything you read and become a more critical reader. Then nobody can bamboozle you with rhetoric.

2. **Help other writers.** You become a valuable critic for other peoples' writing. Using the Writing System and its techniques, you can find specific shortcomings in a document, but more importantly, you can show how to remedy them. Many people can find fault: few can help others improve. Don't be shy. People usually welcome criticism when it accompanies straight-forward remedies.

3. **Speak better.** You discover that the Pre-writing Phase techniques also help you prepare oral presentations.

Appendix B

Quick Reference Guide

Pre-writing Phase

1. Analyze Purpose   1.1 List and contrast three purposes associated with your writing task.

2. Analyze Audience   2.1 List your audiences in priority order.
       2.2 Profile each audience using a checklist.

3. Write a Purpose Statement   3.1 Fill in the five parts for a purpose statement.
       3.2 Use the purpose statement to focus yourself and others.

4. Gather Information   4.1 Use your purpose statement as a guide as you write down ideas.
       4.2 Ask *Who, What, Where, When, Why,* and *How.*

5. Write Sentence Outline   5.1 Write your ideas as assertions.
       5.2 Evaluate your assertions against your purpose statement.
       5.3 Put your assertions in effective order.

Writing Phase

6. Write the Draft   6.1 Sprint through your draft.
       6.2 Identify the source of your writer's block and apply remedy.

Post-writing Phase

7. Revise Content and Organization   7.1 Apply the three-part Content Test.
       7.2 Apply the three-part Organization Test.

8. Edit for Coherence   8.1 Use key words in titles, subheads, and throughout your document.
       8.2 Group your ideas in paragraphs and vertical lists.
       8.3 Preview your document with an introduction.
       8.4 Use front and back matter to help secondary and tertiary audiences.
       8.5 Apply visual devices to help your reader skim, follow, and refer.

9. Edit for Clarity   9.1 Use concrete and specific words.
       9.2 Make verbs active voice, present tense, indicative or imperative mood.
       9.3 Identify and replace ambiguous pronouns.
       9.4 Use standard English words.
       9.5 Be positive.
       9.6 Remove gender bias using nine guidelines.
       9.7 Make sentences parallel.
       9.8 Test modifiers.

10. Edit for Economy   10.1 Cut empty verbs.
       10.2 Cut unnecessary prepositions.
       10.3 Cut who, which, that, and there.
       10.4 Cut repetition.
       10.5 Cut redundancy.
       10.6 Cut implied phrases.
       10.7 Cut unnecessary or vague modifiers.

11. Check for Readability   11.1 Measure readability.
       11.2 Replace long words with short words.
       11.3 Break long sentences.

12. Check for Correctness   12.1 Check word choice.
       12.2 Check grammar.
       12.3 Check punctuation.
       12.4 Check mechanics.

13. Proofread   13.1 Proofread in a series of readings.
       13.2 Check live copy against the dead copy.

Appendix C

Apply System to Long Documents

This appendix expands upon *Step 5 — Sentence Outline* and *Step 8 — Coherence* techniques to help you write longer, more complex technical documents.

The kinds of writing projects that we discuss in this Appendix include Proposals, Requirements Documents, Functional Descriptions, General and Detailed Designs, and manuals. This appendix also helps authors of hypertext management language (HTML) and multi-media applications.

This appendix begins by showing you what a modular layout looks like. Then you overview the techniques needed to master the modular layout. You learn to add the "Storyboard" techniques to your sentence outlining techniques. Storyboards help you organize massive subjects into discrete topics. Also, you learn new coherence techniques to use the modular layout to display discrete topics.

As we show you techniques for writing long documents, we discuss, when appropriate, separate tips for the long proposal document. Proposals are more difficult to write than most long documents, because you need to superimpose "themes" — that is, your competitive advantage. Also, you need to frame your discussions in terms of "benefits," when often the client merely states the request in terms of "features" they want.

Large writing projects overwhelm many writing teams. Your team may find itself confronted with a 400-page deliverable that must address hundreds of complex topics. The team must know where to begin and how to manage time and resources.

The techniques in this appendix help you break the large writing project into manageable tasks. Project Managers and Proposal Team leaders also study Appendix D to learn how to apply proven project management techniques to writing projects.

As a pre-requisite to using these advanced techniques, you need to understand the fundamentals of the Writing System in Chapters 1 though 13. Throughout this appendix, we refer to other Writing System techniques.

Modular layout helps you re-use documents by making topics easy to identify. Topics gracefully fall into other documents that use modular layout. You re-use modular documents in much the same way systems engineers re-use modular software. In systems engineering, we often take modules from one system that we modify to meet the client's specific needs. Similarly, we rarely re-use a whole document; rather, we re-use topics.

Modular layout differs radically from conventional layout. Conventional layout is a continuous flow from topic to topic; topic lengths vary, and the shifts occur anywhere on the page. A modular layout stops at each topic. Each topic occupies two pages (one page each in some cases), and all the topical shifts occur at the top-left corner of the page.

The Writing System Workbook is an example of modular layout. Turn to page 10 and note that the topic is a two-page, left-right presentation. When you turn the page, you notice that you are on a new topic. In fact, every topic begins at the top of the left page and ends on the bottom of the right page. Using this book as a template, you have a typical modular layout where each section has

Pg#	Right	Introduce section
Pg#-pg#	Left-Right	Topic One
Pg#-pg#	Left-Right	Topic Two
Pg#-pg#	Left-Right	Topic Three
Pg#-pg#	Left-Right	Topic n . . .
Pg#	Left	Conclude section

A typical left-right presentation accommodates 960 words, less whatever space you allocate for a figure. Any left over space becomes white space. The modular layout is flexible.

In *The Writing System Workbook*, we use the left page for instructions and right page for exercises. At the top left, the reader sees the subhead that defines the topic. The first sentence that follows is the main assertion or covering generality. Thereafter, the document uses the ordinary coherence devises, such as subheads, paragraphs, and vertical lists, to provide subpoints and supporting details. Typically, the reader finds the graphic, if needed, on the right page. Every graphic has a caption that briefly describes what the graphic does.

The two-page, 960-word limit per topic may seem restrictive, but the limit actually conforms nicely to a typical topic's length. Studies of topic length in conventional documents show that topics vary from 250 to 1,100 words, with a mean around 700 words.

Modular layout in business and technical documents uses about the same amount of white space as does conventional layout.

For a good discussion of why the modular approach works and for a good history of the storyboarding, read *Sequential Thematic Organization of Publications*: STOP, Hughes Aircraft Company, Tracey et al; Fulleton, California; 1965.

These business and technical documents lend themselves naturally to modular layout:

1. proposals
2. requirements
3. functional descriptions
4. detailed designs
5. specifications
6. test plans
7. users manuals
8. procedures and policy documents

Any subject matter that you can break down into discrete topics is a candidate for modular layout. For example, manuals and planning documents naturally break into steps or processes, which you can gracefully display in modular layout. Modular layout was originally developed for writing proposals.

Below are examples of modular layouts.

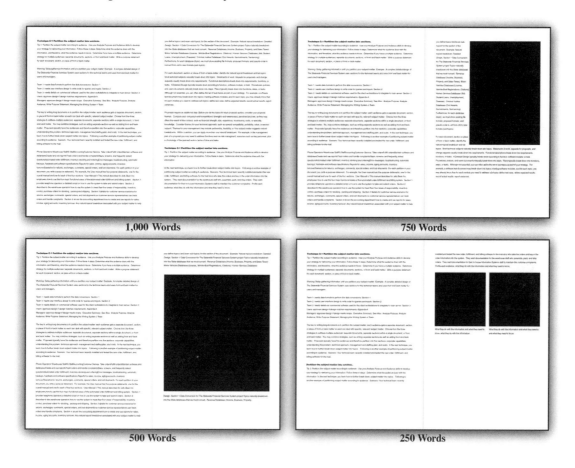

1,000 Words

750 Words

500 Words

250 Words

This overview introduces six techniques to produce a modular layout document. A detailed explanation of each technique follows this overview.

1. Partition subject matter into sections:
Use purpose and audience techniques.
Identify sections according to audience needs.

Phone Operators	Warehouse Staff	IS Staff	Accounting	Customer Service
Take orders	Fulfill orders	Maintain software and databases	Create and use reports	Track orders and handle complaints
Steps, screens, and frequently asked questions	Automated order fulfillment; inventory stocking and ordering	Error messages, troubleshooting, automatic backups, hardware and software specifications	Reports for sales, income, aging accounts, inventory turnover	Scenarios for returns, exchanges comments, special orders, and lost shipments

Write purpose statement for each section.

Section 3 describes how the finance and accounting department creates reports.

2. Break down sections into discrete topics:
Consider what audience needs to know.
Consult predecessor documents.

Users Manual for Jelly Bean Inc.			
Section 1 Telephone Orders	**Section 2 Warehouse Operations**	**Section 3 Customer Service**	**Section 4 Accounting Reports**
Ordinal	Functional	Examples; Steps	Topical
Workstation setup	Taking inventory	Tracking shipments	Daily inventory
Call initiation	Ordering EOQ	Refunds	Inventory turnover
Get customer ID	Receiving orders	Price guarantee rebate	Invoices
Build customer record	Packaging & labeling	Item availability	Aging accounts
Take order	Bundling shipments	Quantity discounts	Cashflow
Methods of payment	Special orders	Complaints to refer	
Total charges	Returns		
Shipping instructions	Safety		
Submit order			

3. Fill in a storyboard for each topic:
Identify topic title.
Make main assertion.
Add supporting assertions.
Add supporting detail.
Sketch graphics.

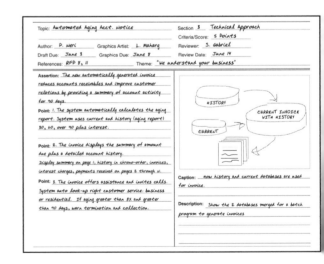

4. Revise and review storyboards:
Check topic titles.
Use Content & Organization Tests.
Check assertions.
Check supporting detail.
Break or combine topics if necessary.

5. Write drafts and produce graphics for
Topics (the body)
Conclusions (each section)
Introductions (each section)

Automated Aging Account Notice

The new automatically generated invoice reduces accounts receivables and improves customer relations by providing them a summary of account activity for 90 days.

Jelly Bean sends invoices on a monthly or bi-weekly billing cycle to different classes of customer: mostly business or residential. The system already has a record of each transaction.

The system automatically calculates the aging report. At the monthly billing cycle, the system automatically sorts the current billing history, then compares that history to the historical record to identify any customers with an outstanding balance due. If the customer has a balance due, the system queries the historical record to extract the information required for the aging report: all invoices more than 30, 60, or 90 days older than the current billing due date. The system calculates interest to charge for each past invoice.

The invoice displays the summary of amount due plus a detailed account history. On page 1, we add a line for past due charges, which is the summary of past due and interest calculated above. If the invoice has a value for

46

past due greater than zero, the printing routine creates as many pages as necessary to display the account history – usually just a second page. The account

Fig. 2 Scheme to add aging report

history shows a table, in chronological order, all invoices, interest charges, and payments with a running balance in a right column.

We propose to add a second table to anticipate customer questions and help customer service answer billing questions. The system automatically presents a second table with all the customer purchases not yet paid in full. Each purchase has the item code, short description, date, quantity, amount, tax, shipping, and employee code of the Fig.2 Scheme to add aging report order taker.

The machine time to generate the table is negligible; the table requires an additional sheet of paper only .01% of the time, so printing and mailing costs are likewise negligible. Meanwhile, your clients get more complete information, are less likely to call your 1-800 number and expend time and money asking questions.

47

Finally, the aging report on the invoice offers assistance and invites calls. The system automatically looks up the appropriate customer service representative name and phone number from a tag in the costumer is current record that directs a quick look-up in the Jelly Bean corporate directory. The invoice printing module prints the suggestion that the customer call his or her service representative with questions. Consequently, your customers – especially your bigger business accounts – resolve problems quickly with less frustration.
If past due account is greater than $X or older than Y days – you can determine both variables for each customer or set as a policy – the invoice printing module prints a warning that you may terminate the account and submit the past due amount to a collection agency.

6. Edit the drafts:
Achieve one voice.
Cut deadwood to fit two-page limit.

Automated Aging Account Notice

The new automatically generated invoice reduces accounts receivables and improves customer relations by providing a summary of account activity for 90 days.

Jelly Bean sends invoices monthly or bi-weekly to different classes of customer: mostly business or residential. The system already has a record of each transaction.

The system automatically calculates the aging report. At the monthly billing cycle, the system sorts current billing history, then compares that history to the historical record to identify customers with a balance due. Then the system extracts information required for the aging report: invoices older than 30 days. Then the system calculates interest charges.

The invoice displays a summary plus a detailed history. On page 1, we add past due charges including interest. If the past due amount exceeds zero, the printing routine creates pages to display account history. An account history shows the order of all invoices, interest charges, and payments with a running balance.

We add a second table to anticipate customer questions and help customer service answer billing questions. The

46

system presents a second table with unpaid purchases. Each purchase has the item code, short description, date, quantity, amount, tax, shipping, and employee code.

Fig. 2 Scheme to add aging report

The time to generate the table is negligible. Likewise printing and mailing costs are negligible. Meanwhile, your clients get more complete information, are less likely to call your 1-800 number and spend time asking questions. Finally, the aging report on the invoice offers help. The system looks up the clients service representative name and phone number from a tag in the client is records and Jelly Bean corporate directory.

The invoice printing module prints the suggestion that the customer call his or her service representative with questions. Consequently, your customers – especially your bigger business accounts – resolve problems quickly with less frustration. If past due account is greater than $X or older than Y days – you set the variables X and Y for each customer – the invoice printing module prints a warning that you may terminate the account and submit the past due amount to a collection agency.

47

Tip 1   Partition the subject matter according to audience. Use your Analyze
Purpose and Audience skills to develop your strategy for delivering your
information. Follow these 4 steps:

1. Determine what the audience does with the information, and therefore,
   what the audience needs to know.
2. Determine if you have a multiple audience.
3. Determine strategy for multiple audiences: separate documents, sections,
   or front and back matter.
4. Write a purpose statement for each document, section, or piece of front or
   back matter.

Warning   Delay gathering information until you partition your subject matter.

Example   A complex detailed design of The Statewide Financial Services System uses
sections for the technical teams and uses front and back matter for users and
managers.

Team 1:   needs data formats to perform the data conversions: Section 1

Team 2:   needs user interface design to write code for queries and inputs: Section 2

Team 3:   needs details on commercial software used for the client workstations to
             integrate to main server: Section 3

Users: approve design if design matches requirements: Appendix A

Managers: approve design if design meets scope: Executive Summary

See also   Analyze Purpose; Analyze Audience; Write Purpose Statement

---

The key to writing long documents is to partition the subject matter: each audience gets a
separate document, section, or piece of front or back matter so each can deal with specific,
relevant subject matter.

Choose from the three strategies to address multiple audiences: separate documents, separate
sections within a single document, or front and back matter. You may combine strategies, such
as writing separate sections as well as adding front and back matter.

Proposals typically have five audiences and therefore partition into five sections: corporate
capabilities; understanding the problem; technical approach; management and staffing plan; and
costs.

In the next technique, you learn how to further break down subject matter into topics.

Following is another example of partitioning subject matter according to audience.

Scenario: Your technical team recently installed and tested the new order, fulfillment, and billing software for the mail order department of Jelly Bean, Inc. Now the company wants you to document the system. Specifically, they want a document for the telephone operators who take the orders and key in the order information into the system. They want documentation for the warehouse staff who assemble, pack, and ship orders. They want documentation for their in-house Information Systems staff to maintain the software and databases. Management wants documentation to explain the report-generating module of the software, primarily for the accounting department. Lastly, the client wants you to document the system for customer service to help them track shipments and handle customer complaints.

Profile each audience: what they do with the information and what they need to know.

Phone Operators	Warehouse Staff	IS Staff	Accounting	Customer Service
Take orders	Fulfill orders	Maintain software and databases	Create and use reports	Track orders and handle complaints
Steps, screens, and frequently asked questions	Automated order fulfillment; inventory stocking and ordering	Error messages, troubleshooting, automatic backups, hardware and software specifications	Reports for sales, income, aging accounts, inventory turnover	Scenarios for returns, exchanges, comments, special orders, and lost shipments

After partitioning the subject matter, you develop a strategy to partition your document. You partition the subject matter into two documents, each with sections or back matter.

Users Manual
Section 1 Telephone Order-taking
Section 2 Warehouse Operations
Section 3 Customer Service
Section 4 Accounting Reports

Maintenance Manual
Section 1 Software Maintenance
Section 2 Database Maintenance
Appendix A Software Specifications
Appendix B Hardware Specifications

For each partition in your document, you write a purpose statement. For example, the User manual has five purpose statements, one for the overall manual and one for each of the four sections.

User Manual — This manual describes for Jelly Bean Inc. employees how to use the four major functional areas of the automated order-fulfillment and billing system.

Section 1 provides telephone operators a detailed script on how to use the system to take and submit orders.

Section 2 describes to the warehouse operators how to use the system to meet their four areas of responsibility: inventory controls, purchase orders for stocking, packing, and shipping.

Section 3 details for customer service scenarios for returns, exchanges, comments, special orders, and lost shipments so customer service representatives can track orders and handle complaints.

Section 4 shows the accounting department how to create and use reports for sales, income, aging accounts, and inventory turnover.

Tip 1   Consider what topics the audience needs to know about in the section.

Use any natural topical breakdown associated with your subject matter to help you define topics (and even sub-topics) for the section.

Warning   Do not gather information about the topics yet. Do not worry about putting the topics in order yet.

Example   Natural topical breakdown: Detailed Design, Section 1—*Data Conversion for the Statewide Financial Services System project*

Topics naturally breakdown into the State databases that we must convert:
Revenue Databases (Income, Business, Property, and Sales Taxes)
Motor Vehicles Databases (License, Vehicle-Boat Registrations, Citations)
Human Services Databases (Aid, Student Loans, Unemployment, Diseases)
Criminal Justice Databases (Civil Awards, Garnishments, Sentencing)

Furthermore, for each database (topic), we must show existing file formats, proposed formats, and logic to convert from old to new formats (sub-topics).

See also   Audience

For each document, section, or piece of front or back matter, identify the natural topical breakdown and list topics.

Most technical subjects naturally break down into topics. Statements of work, requests for proposals, and change requests usually break down into requirements. Functional descriptions break down into requirements, functions, or tasks. A Detailed Design typically breaks down according to function, software module, or task. Procedures, policies, and user documents naturally break down into steps. Plans typically break down into functions, steps, or tasks.

Although not essential, you can often define the set of sub-topics as part of your strategy. For example, a software test document may break down into topics — testing software modules. For each topic, you may already know that for each module you need to address sub-topics: define test case, define expected results, record actual results, report variances.

Proposals require an additional step. Before you list the topics for each proposal section, consider your proposal themes. Compare your company's and competitions' strengths and weaknesses, perceived and real, as they may affect the award of the contract, such as financial strength, size, experience, incumbency, costs, or specialty knowledge. Consider themes for your technical approach, such as upward compatibility, portability, value, or service support. Themes for your management may include partnership, flexibility, or risk mitigation. Devise a list of themes for your proposal, themes that emphasize your strengths and compensate for any perceived weaknesses. Your themes need to answer the question: *why hire us?* Make sure your topics can address your themes.

Subject matter suggests natural breakdowns. Within a section, you can apply more than one natural breakdown. For example, in the management plan of a proposal you may need to address *functions* such as risk management, *resources* such as key personnel, plus a *chronology* of the planned work in terms of time and tasks.

Typical Documents with Sections	**Natural Topical Breakdowns**
Detailed Design	
Section 1 Data Conversion	Database, Step by step
Section 2 Processes	Functions
Section 3 Workstations	Hardware, Software, Modifications
Users Manual	
Section 1 Telephone Order Taking	Step by step
Section 2 Warehouse Operations	Function
Section 3 Customer Service	Situations (topical)
Section 4 Accounting Reports	Reports with examples
Maintenance Manual	
Section 1 Software maintenance	Modules
Section 2 Database Maintenance	Functions
Appendix A Software Specifications	Modules
Appendix B Hardware Specification	Devices
Proposal	
Section 1: Corporate Capabilities	Attributes
Section 2: Understanding the Client Problem	Requirements — general to specific
Section 3: Technical Approach	Process, examples
Section 4: Management Plan	Functions, Resources, Time, Task
Section 5: Cost	Category, Time

Often the Request for Proposal (RFP) dictates the sections and their topics. Otherwise, use your themes to help you determine what natural patterns you choose. Suppose, for example, a major theme is that your company has experience integrating commercial software packages on workstations to large data servers. Your Section 3 Technical Approach ought to show examples of your success to illustrate your approach. On the other hand, you want to stress the theme that you understand the client's *unique* requirements. Your Section 2 Understanding the Client Problem ought to examine the requirements, citing first the *general* industry standard requirements, then focusing on the client's *specific* — perhaps *unique* from their point of view — needs.

Tip 2   Use predecessor documents to determine what the reader needs to know.

If the predecessor document uses a modular layout, you merely extract the discrete topics that need further discussion in your new document. For example, you extract the topics from a Requirements Document to help you determine what the reader needs to know for a Functional Description.

If the predecessor document uses a continuous flow layout, you must deconstruct the document to extract topics. If the document is well organized, extracting topics is as easy as finding the logical breaks as they fall randomly on the page. If the document is poorly organized, you must examine each paragraph to find the point, then regroup those points into topics. (See Sentence Outlining, Technique 5.6)

Use the Edit for Clarity techniques 9.1 through 9.7 to identify ambiguity in your predecessor documents. Resolve questions before you gather information.

Warning   Do not assume that predecessor documents provide the standard for your document.

See also   Sentence Outlining; Clarity

In technical writing, a document often has a predecessor. A Requirements Document may beget a Statement of Work (SOW) or Request for Proposal (RFP), which in turn begets the Proposal. These predecessors often beget a Functional Description that begets a Detailed Design that begets the Test Plan, System Documentation, and the User Manual. One major benefit of using the modular layout is that each successor document becomes easier to write when the predecessor documents already present discrete topics.

For proposal writing, you typically work from an RFP that includes Instructions to the Offerors, Evaluation Criteria, Contract Data Requirements List, Statement of Work, or Work Breakdown Structure. One way or the other, the predecessor document describes the deliverables or services required. Extract from the predecessors a list of compliance criteria and list them in a matrix, referencing the RFP Page and paragraph plus a short description of the criteria. The compliance criteria require some response, and therefore they suggest topics.

So, for proposal writing, you create a detailed list of compliance criteria, which become your topics. You already have a list of themes. Post these two lists in your proposal room.

With your list of topics, you can estimate the length of your document. Each section needs an introduction page (right), a series of topics (left-right), and a conclusion page (left). Therefore, you estimate the number of pages as the (# topics plus 1) times 2. This page-length estimate is a valuable input when estimating the time required to write a document.

The following passage from a Functional Description is in a continuous flow layout and poorly organized at the paragraph level. You can't easily extract topics.

The state requires that the system be able to sort by account number, tax number, date, and account balances. The account balance sorts must include logic that allows the sort by amount of overpayment or underpayment, plus length of time past due. Interest on past due accounts is automatically calculated and compounded daily. Therefore, any past due accounts, must be automatically sorted each day; the new interest and penalty calculations made and the new due amount recoded in the database.

The State Internal Revenue (SIR) system has three databases operating on a Cybase system: Income Tax Revenue, Business Tax, and Property Tax. The State Sales Tax database operates on a standalone relational database system tied directly to the U.S. Department of the Treasury. The state can track both federal and state revenue from the taxes on gasoline.

The databases operating on the Cybase system have the following five fields in common: payer tax number, last name, address, phone number, account balance. In addition the different databases have between 32 and 89 additional fields. The following four fields, although not named identically, are in fact the same.

We may treat them as fields in common upon changing the names to a common name in the SIR:

Income	Property	Business	SIR
FY	Fiscal Year	Acct Year	Fiscal Year
Credit	Overpaid	Refund Due	Credit
Center	Filing Center	Agency	Center

Aside from the commonality stated above, each of the databases has its unique fields. Some of those fields need to be changed to allow easy sorting. For example, the Business Tax database has a name for both the business (sole proprietorship, partnership, or corporation) and the name of the tax payer, which can be a number of entities– an S Corporation passes the tax liability to the shareholders.

The income tax database is by far the largest database with the most relational links to other state and federal agencies. The SIR format uses the Income Tax Database format as much as possible, although all fields must change to plain language instead of acronyms. The fields relate to three general areas: Personal Information, Account History, and Account status. We discuss the fields of each area in turn.

46
47

Note the ease in extracting topics from this modular layout of the same subject matter.

Income Tax Database Layout

The Income tax database layout is the closest to the proposed State Internal Revenue (SIR) layout. Therefore the SIR uses the Income Tax Database format as much as possible, which is by far the largest database with the most relational links to other state and federal agencies.

Commonality

The databases operating on the Cybase system have the following five fields in common: payer tax number, last name, address, phone number, account balance. In addition the Income Tax database has 32 additional fields. We anticipate an increase in the Zipcode field size requirement from 9 to 12 numbers. Some fields need only name changes to plain language instead of acronyms. Therefore, we change field "FY" name only to "Fiscal Year."

Sorting Requirements

The state requires that the SIR sort by account number, tax number, date, and account balances.

The account balance sorts must include logic that allows the SIR to sort by amount of overpayment or under-payment, plus length of time past due. Interest on past due accounts is automatcially calculated and compounded daily. Therefore, the system must automatically sort any past due accounts each day, make the new interest and penalty calculations, and record the new due amount in the Income Tax database.

The following is a table of the 37 fields in the Income Tax Database showing the present version and the SIR version.

46
47

Tip 3   Put the topics in order.

Use natural patterns of thought.

Warning   Do not assume that the topics must flow from one to the next. Many technical documents such as Requirements Documents and Proposals have no chronology or particular ranking. One problem with continuous flow layout is that the flow presupposes a link from topic to topic, when in fact, each topic is independent.

Example

Traditional continuous flow organization purposefully or inadvertently shows subordination of ideas. Although this type of organization is good for essays, the continuous flow organization fails for technical documents that are more topical. In contrast, the modular organization simply lists the major topics. In this example, the continuous flow infers a relationship of ideas that does not really exist and is therefore confusing. The modular organization avoids giving the false relationships by merely listing the major topics.

Continuous Flow Organization

I. Database Conversion Requirements
   A. Introduction
      1. Scope of the Changes
      2. Two Methods for Conversions
         a. Rename Fields and Copy
         b. Conversion Programs
   B. Common fields
      1. Income Tax Revenue Database
      2. Other Tax Revenue Databases
         a. Property Taxes
            (1) Automobiles
            (2) House and Furnishings
         b. Business taxes
      3. Sales Taxes

Modular Organization

I State Revenue Database Basics
II Conversion of Income Tax Database
III Conversion of Property Tax Databases
IV Conversion of Sales Tax Database

See Also   Sentence Outline; Natural Patterns of Thought

When you write your topics, use descriptive language. Instead using of just a noun, add a verb or prepositional phrase to convey meaning.

For example
Poor Topic: Income Tax Database
Better: Conversion of Income Tax Database

Continuous flow organizations tend to be vertical and hard to follow. Modular or topical organization is more horizontal and easier to follow.

Continue with the Jelly Bean, Inc. scenario. You already partitioned the subject matter into two documents with sections plus front and back matter. Now you break down each partition into discrete topics. The charts below also indicate the natural pattern of thought employed.

Users Manual for Jelly Bean Inc.			
Section 1 Telephone Orders	**Section 2 Warehouse Operations**	**Section 3 Customer Service**	**Section 4 Accounting Reports**
Ordinal	*Functional*	*Examples; Steps*	*Topical*
Workstation setup	Taking inventory	Tracking shipments	Daily inventory
Call initiation	Ordering EOQ	Refunds	Inventory turnover
Get customer ID	Receiving orders	Price guarantee rebate	Invoices
Build customer record	Packaging & labeling	Item availability	Aging accounts
Take order	Bundling shipments	Quantity discounts	Cashflow
Methods of payment	Special orders	Complaints to refer	
Total charges	Returns		
Shipping instructions	Safety		
Submit order			

Maintenance Manual for Jelly Bean Inc.			
Section 1 Software Maintenance	**Section 2 Database Maintenance**	**Appendix A Software Specifications**	**Appendix B Hardware Specifications**
Topical; cause effect	*Ordinal*	*Functional*	*Spatial, topical*
Change Management Process	Daily string search	GUI Interface	Parallel Processor Server
Communications Security tests	Data contamination tests	Zybert dBase	ISD Communications Suite
Nightly error log	Nightly backup	YYServer Com	Workstation – Customer service
	Weekly merge-purge	Processes	Workstation – Warehouse
	Weekly "fire" copy	Lotus Notes™ with modifications	LAN server
	Monthly archival		

Tip 1   Design your storyboard form.

Fill in the overhead beginning with a good title for your topic.

Warning   Do not add overhead to the storyboard sheet unless you know you need the information.

Example   Design the overhead for a proposal storyboard to keep track of assignments, deadlines, compliance criteria, scoring, references and themes.

Topic: *Automated Aging Acct. Notice*   Section *3*   *Technical Approach*

Criteria/Score: *5 Points*

Author: *P. Neri*   Graphics Artist: *L. Maharg*   Reviewer: *S. Gabriel*

Draft Due: *June 3*   Graphics Due: *June 8*   Review Date: *June 14*

References: *RFP 8, 11*   Theme: *"We understand your business"*

See also   Sentence Outlining

Design your own storyboard blank sheets on 8½"x11" or 11"x17" paper. You can adapt from the formats pictured in this appendix, but build a format to suit your writing project. Limit the overhead items to those you need.

Typical overhead items include
        topic title
        responsibilities such as author, graphics artist, reviewer, editor
        due dates for pink team review, draft, graphics, final copy, production
        predecessor documents such as specific items on the deliverables and criteria list
        key linkage to other sections

Proposals typically need more overhead. Add themes, compliance criteria, and scoring to the overhead. Often, an RFP tells you how the reviewers intend to score the different compliance criteria, and you want to alert authors and reviewers to distinguish between big and little issues.

Under the Overhead, the storyboard blank needs space for
Left side   Topic   Right Side   Sketch of graphic
               Main assertion   Caption
               Subordinate points   Word description of graphic

After you develop the whole set of storyboards, you can review and revise the overhead as well as the topic headings, assertions, and supporting details.

If office space allows, put your storyboards on a wall where everyone on the team can see them.

Following is a design of a storyboard for a Technical Approach section of a proposal for the topic *automated aging account notice*: Predecessor document is the Request for Proposal pages 8 and 11.

Topic: **Automated Aging Acct. Notice**	Section **3** **Technical Approach**
	Criteria/Score: **5 Points**
Author: **P. Neri** Graphics Artist: **L. Maharg**	Reviewer: **S. Gabriel**
Draft Due: **June 3** Graphics Due: **June 8**	Review Date: **June 14**
References: **RFP 8, 11**	Theme: **"We understand your business"**

Assertion: _____

_____

_____

Point: _____

_____

_____

Point: _____

_____

_____

Point: _____

_____

_____

Point: _____

_____

_____

Caption: _____

_____

Description: _____

_____

_____

_____

If you can make 11"x17" storyboards, and if your storyboard room can display that much paper, the large size helps keep your work from getting cramped.

Tip 2   Write assertions or covering generalities for each topic.

Write 2 to 5 minor points that support your main assertion.

Warning   Do not simply repeat the assertions from a predecessor document.

Example

The RFP states: "BAPCO requires that the new fulfillment system operate 500% faster than the current system, and the system must be up-gradable to operate on larger machines so we can handle larger databases in the future without loss of speed."

Too often, the proposal writer merely repeats the words used in the RFP, coming up with weak assertions:

- We propose to build a system that operates at least 500% faster than current systems.
- Our proposed architecture allows you to upgrade to faster machines to handle larger databases.

The clever proposal writer makes inferences. In the example above, the client wants the speed to increase throughput per employee to achieve a higher return on sales. The client wants to upgrade their system because they anticipate growth. Therefore, the stronger assertions are

- We propose to increase your return on sales by giving your sales force a 500% faster fulfillment system.
- We shall build a system that grows with your business.

See also   Sentence Outlining

Technical documents such as requirements, functional descriptions, detailed designs, and manuals have mostly covering generalities. Make your generalities *results oriented*. Instead of "Begin with these three steps," write "Log into your personal mailbox by following these three steps."

When writing assertions for proposals, take your list of compliance criteria (See C-1, Tip 2) and infer a benefit for each criterion. Then write your assertions based on the inferred benefits — do not simply repeat the criteria or requirements. A typical RFP stated: *"Simply rephrasing or restating the Government's requirements is insufficient and will result in the proposal being considered non-responsive." — DTFA01-89-R-00215.* To write winning assertions, you must infer the benefits that the client seeks.

Also for proposals, consult your list of themes as you craft winning assertions. Your themes often speak to the benefits of hiring your firm. For example: *As the incumbent vendor who helped you analyze your business processes, we are uniquely suited to build your automated order-fulfillment system.*

In the following storyboard for a proposal, we take the *feature* requested in an RFP and evolve it into a *benefit* that we express as an *assertion*.

Feature: The billing system must print invoices that provide a summary of account activity for the past 90 days.

Benefit: The account summary helps collections by casually reminding the customers of past due accounts, and by helping the customer and the service representative quickly resolve any disputed charges.

Assertion: The new automatically generated invoice reduces accounts receivables and improves customer relations by providing a summary of account activity for 90 days.

Write your assertion plus minor points on your storyboard:

Topic: _Automated Aging Acct. Notice_	Section _3_ _Technical Approach_
	Criteria/Score: _5 Points_
Author: _P. Neri_ Graphics Artist: _L. Maharq_	Reviewer: _S. Gabriel_
Draft Due: _June 3_ Graphics Due: _June 8_	Review Date: _June 14_
References: _RFP 8, 11_ Theme: _"We understand your business"_	

Assertion: _The new automatically generated invoice reduces accounts receivables and improve customer relations by providing a summary of account activity for 90 days._

Point: _1. The system automatically calculates the aging report._

Point: _2. The invoice displays the summary of amount due plus a detailed account history._

Point: _3. The invoice offers assistance and invites calls._

Caption: _____

Description: _____

Work at the level of assertion and minor point for all your topics, *before* you add detail or draw sketches to any of your topics.

Tip 3   Add supporting details and sketch the graphic.

Without writing the paragraphs or vertical lists, jot down the key supporting details for your major assertion and minor points.

Sketch the graphic, table or figure in minimal detail. Write the caption for the graphic as specifically as possible. Write a note to the graphic artist describing what you want the graphic to do.

Warning   Do not concern yourself with *how you say it;* rather, focus on *what you say.* Before you write the draft, time spent worrying about word choice, tone, and phrasing is usually wasted.

Example   Detailed Design storyboard topic, Conversion of Income Tax Databases

Assertion: To convert the Income Tax Database into SIR layout, use Cybase utilities plus our proprietary SIR conversion program.

Minor Point 1 Rename four fields with the Cybase utility: RENAME.
Supporting Details: change add_2 to add_b; FY to fiscal year; Zipcode from 9 to 12 numbers; change Zipcode from alpha to numeric; re-sort.
Minor Point 2. Use the following field names, field lengths, and other field characteristics to run the SIR conversion program.
Supporting Details: Turn off "Check for Duplicate records." See table below for values.

Field Name	Field Length	Number-Character	Logical	Comments
Last_Name	50	Characters	N	Hot key to Business_Name
First_Name	50	Characters	N	None

See also   Sentence Outline

Writing supporting details onto the storyboard is a departure from our sentence outlining technique. Recall that for normal-sized documents, we add the detail when we write the draft. Then we revise content and organization as the first step of editing. However for long documents, we add the key details in the storyboard, then we review and revise *before* we write the draft. Consequently, reviewers need to see supporting details.

Group your details under the points they support. If you have important details that do not fall under an assertion or minor point, you either add the point, or you find the supporting point on another storyboard.

After you (and perhaps your colleagues) add supporting details, expect your storyboards to look messy. In fact, another common name for storyboards is *scribble-boards.*

The following storyboard shows the added detail and sketch.

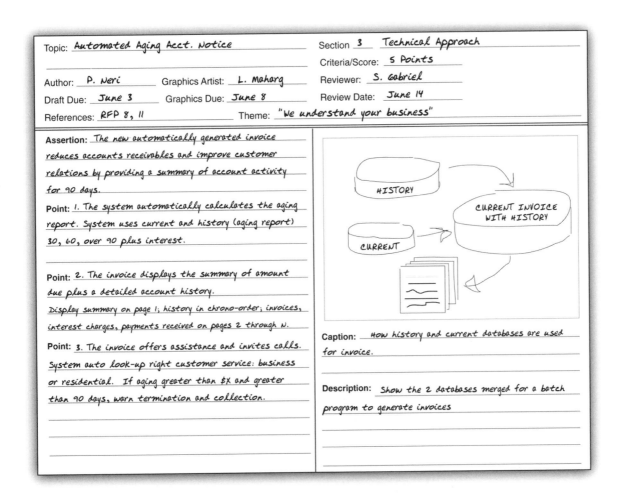

Topic: __Automated Aging Acct. Notice__   Section __3__ __Technical Approach__
                                                 Criteria/Score: __5 Points__

Author: __P. Neri__   Graphics Artist: __L. Maharg__   Reviewer: __S. Gabriel__

Draft Due: __June 3__   Graphics Due: __June 8__   Review Date: __June 14__

References: __RFP 8, 11__   Theme: __"We understand your business"__

Assertion: _The new automatically generated invoice reduces accounts receivables and improve customer relations by providing a summary of account activity for 90 days._

Point: _1. The system automatically calculates the aging report. System uses current and history (aging report) 30, 60, over 90 plus interest._

Point: _2. The invoice displays the summary of amount due plus a detailed account history._
Display summary on page 1; history in chrono-order; invoices, interest charges, payments received on pages 2 through N.

Point: _3. The invoice offers assistance and invites calls. System auto look-up right customer service: business or residential. If aging greater than $X and greater than 90 days, warn termination and collection._

Caption: _How history and current databases are used for invoice._

Description: _Show the 2 databases merged for a batch program to generate invoices_

In the storyboard room, you find different colored pens, tape, stick-on notes, staplers, and paper clips—just about anything you can imagine to mark and past information on a wall. In practice, your storyboard may become littered with stick-on notes, excerpts or boilerplate from other documents, even news clippings, meeting notes, and clarifications. Before you review and revise, you need to tidy your storyboards. Display only the details you intend to use. Save the other odds and ends in a folder just in case the review determines that you need one of those bits of information.

Tips   First, check each storyboard as a separate topic.
Apply the Content and Organization Tests for individual storyboards. (See Chapter 7.)

Specifically check the following five items:
1. Check topic headings. Make them concrete and specific.
2. Check the major and minor points to make sure they address what the reader needs to know.
3. Check details to ensure that they adequately support the points.
4. Check graphics to ensure value; ensure that caption describes the point made in the graphic.
5. Check for modular layout problems. Split a topic that has grown too large, or combine two topics for which you have little to say.

Second, check the storyboards as they relate to each other:
1. Identify and eliminate redundancy.
2. Identify holes in your document.

Warning   Wait until the team finishes all the storyboards. Part of the process is to ensure that you eliminate redundancy and fill in gaps, which you cannot do until you see the whole document in storyboard form.

Do not bother with tone, word choice, grammar, punctuation, or mechanics—or any other editing concerns.

See also   Content and Organization Tests

Invite your graphics professionals to the review. They can suggest better graphics and avoid the waste of trial-and-error graphics production.

If you need to split a long topic, elevate minor points to the level of topic. For example, the following topic fills much more than two pages: *Sending and Receiving E-mail by means of intracompany access to the Internet.* You can break the topics into
Sending E-mail by means of intracompany access to the Internet
Receiving E-mail by means of intracompany access to the Internet

Using reverse logic, you combine two or more short topics into one.
Short topic *Benefits of using graphite fiber in fuel cells* barely fills one page
Short topic *Risks of using graphite fiber in fuel cells* barely exceeds one page.
Combined topic, *Benefits of using graphite fibers in fuel cells outweighs the risks* makes a two-page topic.

In addition to combining and splitting topics, you can occasionally cheat. For example, on pages 114-117 in this book we expanded the exercise from one to three pages, turning our typical two-page topic into a rare four-page topic.

The proposal team, plus other company experts, review and revise the storyboards. Often called the *Pink Team*, they critique the content to make sure

- the proposal addresses all the selection criteria in the RFP
- topics address competitive themes
- assertions are relevant and address benefits instead of features
- supporting details and graphics are relevant and true

In addition, the Pink Team

- ensures that topics can fit the left-right layout (or must be split or combined)
- removes topic heading vagueness
- identifies gaps or redundancies

When you perform the Pink Team for a deliverable, such as a functional description or a user manual, invite the client to participate. You thereby virtually guarantee that your client will accept the document.

Tips   Write a two-page draft for each topic in the section. Write up to 50% more words than you think you need. If you think you need 700 words to cover all the material in a topic, write up to 1,050 words. Feel confident that you can edit those words down to your 700-word limit.

Write the one-page conclusion for each section, usually *what happens next*.

Write the one-page introduction for the section.

Warning   Do not interrupt your writing by editing.

See also   Writing Phase

Transcribe the information from the storyboards into your wordprocessor, but leave the storyboards on the wall as a reference tool for the team.

Because you review and revise the storyboards before you write your draft, you minimize surprises. Nevertheless, if you get a new idea or you have a question, make a note and stick it to the appropriate storyboard when you take a break.

Storyboards and modular layout let you break large writing jobs into manageable tasks. As you write, concentrate on one topic — do not divert your attention to other topics.

Do not worry if you see white space. Studies show that "continuous flow" documents and "modular layout" documents have similar amounts of white space.

Compare your final drafts to the storyboards to ensure that the topics, assertions, minor points, supporting details, graphics and captions convey the message. If you are writing a proposal, make a special pass to ensure that the draft addresses your list of themes, and addresses each point in the evaluations criteria list.

Continuing the example of the proposal, the author writes more draft than two pages can hold.

Automated Aging Account Notice

The new automatically generated invoice reduces accounts receivables and improves customer relations by providing them a summary of account activity for 90 days.

Jelly Bean sends invoices on a monthly or bi-weekly billing cycle to different classes of customer: mostly business or residential. The system already has a record of each transaction.

The system automatically calculates the aging report. At the monthly billing cycle, the system automatically sorts the current billing history, then compares that history to the historical record to identify any customers with an outstanding balance due. If the customer has a balance due, the system queries the historical record to extract the information required for the aging report: all invoices more than 30, 60, or 90 days older than the current billing due date. The system calculates interest to charge for each past invoice.

The invoice displays the summary of amount due plus a detailed account history. On page 1, we add a line for past due charges, which is the summary of past due and interest calculated above. If the invoice has a value for

46

past due greater than zero, the printing routine creates as many pages as necessary to display the account history – usually just a second page. The account

Fig. 2 Scheme to add aging report

history shows a table, in chronological order, all invoices, interest charges, and payments with a running balance in a right column.

We propose to add a second table to anticipate customer questions and help customer service answer billing questions. The system automatically presents a second table with all the customer purchases not yet paid in full. Each purchase has the item code, short description, date, quantity, amount, tax, shipping, and employee code of the Fig.2 Scheme to add aging report order taker.

The machine time to generate the table is negligible; the table requires an additional sheet of paper only .01% of the time, so printing and mailing costs are likewise negligible. Meanwhile, your clients get more complete information, are less likely to call your 1-800 number and expend time and money asking questions.

47

Finally, the aging report on the invoice offers assistance and invites calls. The system automatically looks up the appropriate customer service representative name and phone number from a tag in the client's current record that directs a quick look-up in the Jelly Bean corporate directory. The invoice printing module prints the suggestion that the customer call his or her service representative with questions. Consequently, your customers – especially your bigger business accounts – resolve problems quickly with less frustration.

If past due account is greater than $X or older than Y days – you can determine both variables for each customer or set as a policy – the invoice printing module prints a warning that you may terminate the account and submit the past due amount to a collection agency.

This draft is about 20% too long, but we can shrink the graphic and cut some deadwood to make the text fit.

Tips   Use *The Writing System Workbook* editing techniques to

- give the message one clear meaning
- make topics fit the two-page limit
- provide one voice

Check the entire document — *all sections, front and back matter* — for consistency.

Ensure that the document meets any client-specified standards for word choice, acronyms, symbols, and other conventions.

Warning   Unless you find a glaring deficiency, avoid making changes to the content. In other words, do not let the *perfect* become the enemy of the *good*.

See also   Proofreading

Authors edit their own drafts for clarity and economy, while technical editors, if available, typically add the most value when editing for coherence and readability. Both authors and editors then correct word choice, grammar, punctuation, and mechanics, then proofread for quality assurance.

A coherence device that works well with modular layout is to **Bold** the assertions and minor points.

Before you split a topic, try using deadwood cutting techniques (See Chapter 10) to fit the topic into the two-page limit.

A systematic edit of the document sections gives the whole document one voice, *even if your team of writers come from different cultures and professional backgrounds.* In the end, all the document paragraphs begin with an assertion—a short sentence. During the edit for clarity, everybody cuts passive voice, subjunctive mood, and ambiguous pronouns. The edit for economy eliminates many of the clichés found as prepositions and implied phrases, and thereby rids the document of regional dialect. Finally, the edit for readability makes everybody's prose read at approximately the same grade level.

The following is a picture of the draft after the editing. Deadwood cutting alone reduced the draft enough to fit the text within the two-page limit:

Automated Aging Account Notice

The new automatically generated invoice reduces accounts receivables and improves customer relations by providing a summary of account activity for 90 days.

Jelly Bean sends invoices monthly or bi-weekly to different classes of customer: mostly business or residential. The system already has a record of each transaction.

The system automatically calculates the aging report. At the monthly billing cycle, the system sorts current billing history, then compares that history to the historical record to identify customers with a balance due. Then the system extracts information required for the aging report: invoices older than 30 days. Then the system calculates interest charges.

The invoice displays a summary plus a detailed history. On page 1, we add past due charges including interest. If the past due amount exceeds zero, the printing routine creates pages to display account history. An account history shows the order of all invoices, interest charges, and payments with a running balance.

We add a second table to anticipate customer questions and help customer service answer billing questions. The

system presents a second table with unpaid purchases. Each purchase has the item code, short description, date, quantity, amount, tax, shipping, and employee code.

Fig. 2 Scheme to add aging report

The time to generate the table is negligible. Likewise printing and mailing costs are negligible. Meanwhile, your clients get more complete information, are less likely to call your 1-800 number and spend time asking questions. Finally, the aging report on the invoice offers help. The system looks up the client's service representative name and phone number from a tag in the client's records and Jelly Bean corporate directory.

The invoice printing module prints the suggestion that the customer call his or her service representative with questions. Consequently, your customers – especially your bigger business accounts – resolve problems quickly with less frustration. If past due account is greater than $X or older than Y days – you set the variables X and Y for each customer – the invoice printing module prints a warning that you may terminate the account and submit the past due amount to a collection agency.

46

47

Modular Layout and Storyboarding	Continuous Flow Method
Writing modular documents is easier to manage. Partitioning subjects down to two-page topics scopes your effort. You can more accurately budget and manage time—essential for projects with tight deadlines.	Without a way to scope the size of the document, you have difficulty budgeting and managing time for large document efforts.
Boundaries for each topic ensure that you give appropriate weight to each topic; they keep the author from over-reaching with unnecessary detail. Modular layout helps you write documents with a page limit, because you allocate your allotted pages to topics.	Lack of boundaries lets the author skimp on hard topics and over-write familiar topics. In proposals, this lack of discipline causes you to unwittingly highlight your weaknesses or uncertainties.
Strengthened coherence helps the reader skim the document, follow the logic, and refer back to topics. A modular layout document is easier for your in-house experts to critique and easier for your clients to use. Figures always accompany text.	Continuous flow often implies relationships among topics that don't really exist, thus confusing the reader. The major assertion or covering generality is not prominent, if it exists. Figures do not always accompany text, and thus are difficult to find.
Modular layout reduces production problems: pagination, figures, widows and orphans. Indexing is easier. Modular layout helps when writing "living documents," such as software designs or policy documents that change. The topical left-right presentation lets you "pull out and plug in" material.	Changes tend to have a *ripple effect* throughout your document.
Modular layout helps you re-use documents, because topics are easy to identify: they begin at the top of the left page and end at the bottom of the right page. Because the modular layout organizes your documents into discrete topics, you can easily find and extract those topics you wish to re-use in other documents.	In continuous flow documents, topics begin and end randomly on the pages throughout the document and are therefore difficult to identify.

Index

G

Gather information 23-28

Gender bias 82

Grammar

    See Clarity 67-88

        Economy 89-104

        Readability, break long
        sentences 110

        Correctness 118-125

Grammar-checking software limitations
114

H

Headers and footers 52

Hyphens 136

I

Imperative mood 70, 74

Inconsistency

    See Shifting key words 52

        Mechanics 140

Index

    See Front and back matter 62

Indicative mood 70, 74

Introduction

    See Analyze purpose 6

        Coherence 60

Irrelevancies

    See Sentence outline 32

        Content Test 46

Italics 64

J-L

Jargon

    See Analyze audience 9-14

        Concrete and specific words 68

Justification

    See Coherence 64

        Mechanics 140

Key words 52

Lists (vertical)

    See Coherence 54-57

        Punctuation of lists 138

Lists used to cut repetition 96

M

Mechanics 140

Modifiers

    Dangling or misplaced 86

    Cutting unnecessary 102

Mood (indicative, infinitive, and
subjunctive) 74

N

Natural patterns of thought 36

Negatives

    See Adjective, adverb 124

        Clarity 80

Non-English 78

Noun

    See Concrete and specific words 68

    Used to replace ambiguous
    pronouns 76

Numbers vs. bullets for lists

    See Coherence 56

O

P

R

To Our Readers

Dear Reader:

We welcome your comments and suggestions about *The Writing System Workbook*. Please write, call, or fax your thoughts to

Graham Associates
9117 Saranac Court Fairfax, VA 22032
Telephone: (703) 978-0122 FAX: (703) 978-0525

Email dograham@erols.com

Visit our home page on the Internet at http://www.erols.com/dograham.

Also, if you need more copies of *The Writing System Workbook*, or want information about our writing seminars, please call us directly at (703) 978-0122.

Best regards,

Daniel and Judith Graham